Physics
for the
Logic Stage

Student Guide

Physics for the Logic Stage Student Guide

First Edition, 2nd Printing
Copyright @ Elemental Science, Inc.
Email: support@elementalscience.com

ISBN# 978-1-935614-41-8

Printed in the USA for worldwide distribution

For more copies write to:
Elemental Science
PO Box 79
Niceville, FL 32588
support@elementalscience.com

Copyright Policy

Physics for the Logic Stage
Student Guide Table of Contents

Physics for the Logic Stage
Letter to the Student

Dear Student,

Welcome to your journey through physics, which is the study of forces, motion, and more. This year you will examine the basics of physics, the concepts, and how these principles can be applied. You will look at sound, light, electricity, robots, and more along your voyage. This guide is written to you, so enjoy your journey!

What does this guide contain?

First, this guide includes the Date Sheets and Unit Materials for each of the units. The Unit Materials include the Vocabulary Sheet for the unit, weekly Student Assignment Sheets, blank sketches, Experiment Sheets, and space for each of your writing assignments. After the Unit Materials, you will find the Appendix of this guide. In it you will find a list of all the memory work for the year, a glossary, and a place to record any additional activities you have done that pertain to biology.

Student Assignment Sheets

The Student Assignment Sheets contain your weekly assignments for each week. Each of the student assignment sheets contains the following:

✓ **Experiment** – Each week will revolve around a weekly topic. You will be assigned an experiment to complete that poses a question about what you are studying. Each Student Assignment Sheet contains the list of materials you will need and the instructions to complete the experiment. This guide also includes experiment sheets for you to fill out each week. In each of these experiments, you will use the scientific method.

A Word about the Scientific Method – The scientific method is a method for asking and answering scientific questions. This is done through observation and experimentation. The following steps are key to the scientific method:

1. **Ask a Question** – The scientific method begins with asking a question about something you observe. Your questions must be about something you can measure. Good questions begin with how, what, when, who, which, why, or where.
2. **Do Some Research** – You need to read about the topic from your question so that you can have background knowledge of the topic. This will keep you from repeating mistakes of the past.
3. **Formulate a Hypothesis** – A hypothesis is an educated guess about the answer to your question. Your hypothesis must be easy to measure and answer the original question you asked.
4. **Test with Experimentation** – Your experiment tests whether your hypothesis is true or false. It is important for your test to be fair. This means that you may need to run multiple tests. If you do, be sure to only change one factor at a time so that you can determine which factor is causing the difference.

5. **Record and Analyze Observations or Results** – Once your experiment is complete, you will collect and measure all your data to see if your hypothesis is true or false. Scientists often find that their hypothesis was false. If this is the case, they will formulate a new hypothesis and begin the process again until they are able to answer their question.

6. **Draw a Conclusion** – Once you have analyzed your results, you can make a statement about them. This statement communicates your results to others.

Each of your experiment sheets will begin with a question and an introduction. The introduction will give you some background knowledge for the experiment. The experiment sheet also contains sections for the materials, a hypothesis, a procedure, an observation, and a conclusion. In the materials section, you need to fill out what you used to complete the experiment. In the hypothesis section, you need to predict the answer to the question posed in the lab. In the procedure section, you need to write a step-by-step account of what you did during your experiment. In other words, you need to provide enough detail so that someone else could read your report and replicate your experiment. In the observation section, you need to write what you saw and observed as well as any results you measured. Finally in the conclusion section, you need to write whether or not your hypothesis was correct and any additional information you have learned from the experiment. If your hypothesis was not correct, discuss why with your teacher and then include why your experiment did not work on your experiment sheet.

Safety Advisory—Do not perform any of the experiments marked " ☺ **CAUTION** " on your own. Be sure you have adult supervision.

☐ **Vocabulary and Memory Work** – Throughout the year, you will be assigned vocabulary and memory work for each unit. Each week, you will need to look up the word in the glossary on pp. 271-276 and fill out the definitions on the Unit Vocabulary Sheet found at the beginning of each unit. You may also want to make flash cards to help you work on memorizing these words. Each week, you will also have a memory work selection. Simply repeat this selection until you have it memorized, and then say the selection to your teacher. There is a complete listing of the memory work selections in the Appendix on pp. 261-264.

▦ **Sketch** – Each week, you will be assigned a sketch to complete. Color the sketch and label it with the information given on the Student Assignment Sheet. Be sure to give your sketch a title.

✍ **Writing** – Each week, you will be writing an outline and/or a narrative summary. The student assignment page will give you a reading assignment for the topic from your spine text, either *DK Science Encyclopedia, Bridges and Tunnels,* or *Robotics.* After you have finished the assignment, discuss what you have read with your teacher. Your teacher will let you know whether to write an outline or a narrative summary from your reading. Your teacher may also assign additional research reading out of the following books:
   📖 *The Kingfisher Science Encyclopedia (KSE)*
   📖 *Usborne Illustrated Dictionary of Science (UIDS)*

Once you finish the additional reading, prepare a narrative summary about what you have learned from your reading. Your outlines should be three-level main topic style outlines and your narrative summaries should be three to four paragraphs in length, unless otherwise assigned by your teacher.

- ⏱ **Dates** – Each week, dates of important discoveries within the topic and dates from the readings are given on the student assignment sheet. You will enter these dates onto one of four date sheets. The date sheets are divided into the four time periods laid out in *The Well-Trained Mind* by Susan Wise Bauer and Jessie Wise (Ancients, Medieval-Early Renaissance, Late Renaissance-Early Modern, and Modern). These sheets are found in the ongoing projects section of this guide. You can choose to just write the dates and information on the sheet or you can draw a timeline in the space provided and enter your dates on that.

How to schedule this study

Physics for the Logic Stage is designed to take up to three hours per week. You, along with your teacher, can choose whether to complete the work over five days or over two days. Below are two options for scheduling to give you an idea of how you can schedule your week:

- ✓ A typical two-days-a-week schedule
  - ⏱ **Day 1** – Define the vocabulary, do the experiment, complete the experiment page, and record the dates.
  - ⏱ **Day 2** – Read assigned pages and discuss together, prepare the science report or outline, and complete the sketch.
- ✓ A typical five-days-a-week schedule
  - ⏱ **Day 1** – Do the experiment and complete the experiment page.
  - ⏱ **Day 2** – Record the dates and define the vocabulary.
  - ⏱ **Day 3** – Read assigned pages and discuss together and complete the sketch.
  - ⏱ **Day 4** – Prepare the science report or outline.
  - ⏱ **Day 5** – Complete one of the Want More activities from the Teacher Guide.

Final Thoughts

As the author and publisher of this curriculum, I encourage you to contact me with any questions or problems that you might have concerning *Physics for the Logic Stage* at support@ elementalscience.com. I will be more than happy to answer them as soon as I am able. I hope that you will enjoy *Physics for the Logic Stage*!

Sincerely,
Paige Hudson, BS Biochemistry, Author

Ancients 5000 BC–400 AD

Medieval–Early Renaissance 400AD–1600AD

Late Renaissance-Early Modern 1600 AD-1850 AD

Late Renaissance-Early Modern 1600 AD-1850 AD

Modern 1850 AD–Present

Modern 1850 AD-Present

Physics
Unit 1

Forces and Motion

Unit 1 Forces and Motion
Vocabulary Sheet

Define the following terms as they are assigned on your Student Assignment Sheet.

1. Balance – _____

   _____

   _____

2. Force – _____

   _____

   _____

3. Force field – _____

   _____

   _____

4. Newton – _____

   _____

   _____

5. Air resistance – _____

   _____

   _____

6. Friction – _____

   _____

   _____

7. Gravity – _____

   _____

   _____

8. Terminal velocity – _____

   _____

   _____

9. Inertia – _____

_____

_____

10. Mass – _____

_____

_____

11. Momentum – _____

_____

_____

12. Weight – _____

_____

_____

13. Acceleration – _____

_____

_____

14. Speed – _____

_____

_____

15. Velocity – _____

_____

_____

Student Assignment Sheet Week 1
Forces

Experiment: Can I Measure Force?

Materials

- ✓ Thick, sturdy cardboard
- ✓ 1 Brad fastener
- ✓ Rubber band
- ✓ Hole punch or nail
- ✓ String – about 3 in (10 cm)
- ✓ 3 Jumbo paper clips
- ✓ Pen
- ✓ Objects of varying weight

Procedure

1. Read the introduction to the experiment and then begin to assemble your force meter. Cut out a 3.5 in (9 cm) by 12 in (31 cm) rectangle from the cardboard. Then, punch a hole with the hole punch or nail near the top, large enough for the brad fastener to slide through. Slip one of the paper clips through the brad, through the hole, and fasten the brad on the opposite side. Slide the rubber band onto the opposite end of the paper clip. Next, take another paper clip and turn out a portion of the end to make a pointer. Tie the string to one end of the pointer paper clip and then slide the other end onto the rubber band. Take the third paper clip and fashion a hook out of it. Once you are done, attach the hook to the other end of the string. Hold your force meter at the top and mark where the pointer rests. This line will be your zero force mark. Now draw a scale down the remainder of your force meter. You can use finger widths, inches, or centimeters for your scale, just as long as you use the same measurement for each mark. (**Note**—*You will need your force meter for next week's experiment as well.*)
2. Now that the force meter is assembled, you can use it to measure the different objects. Simply attach each object to the hook and observe what happens. Write down how much the rubber band stretched on the experiment sheet. Repeat this process for each of your objects.
3. Draw conclusions and complete the experiment sheet.

Vocabulary & Memory Work

- ☐ Vocabulary: balance, force, force field, newton
- ☐ Memory Work—This week, work on memorizing the force equation:
  - ⚡ 1 Newton (N) = 1 kilogram (kg) • 1 meter (m) / second (s^2)

Sketch: Resultant Force

- 🗺 Label the following—Forces pull in the same direction; add the forces together to get the resultant force; forces pull in equal, but opposite directions; the forces will cancel each other out for a zero resultant force; forces pull unequal, opposite directions; subtract the forces to get the resultant force.

Writing

- ✍ Reading Assignment: *DK Encyclopedia of Science* pp. 114-115 (Forces), pg. 116 (Combining Forces), and pg. 117 (Balanced Forces)
- ✍ Additional Research Readings
  - 📖 Force: *KSE* pp. 290-291, *UDIS* pp. 6-7

Dates

- 🕐 c330 BC – Aristotle proposes that a force is needed to maintain motion.
- 🕐 1642-1727 – Isaac Newton, the English scientist who explained how force, mass, and acceleration are related, lives. The unit of force, the newton (N), is named after him.
- 🕐 1979 – Pakistani scientist, Abdus Salam, wins the Nobel Prize in Physics for his work with forces.

Sketch Week 1

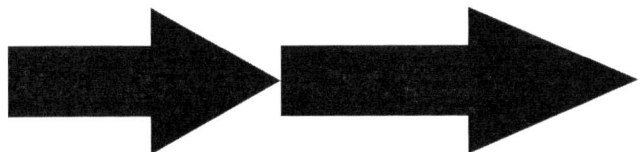

Can I Measure Force?

Introduction

Forces are all around us. They push and pull objects, causing them to move or change shape. In today's experiment, you are going to create your own force meter that can measure the amount of force an object exerts. In a force meter, an object applies a downward force, which stretches a rubber band or spring. We can measure the amount of displacement to determine how much force was applied.

Materials

_____   _____

_____   _____

_____   _____

_____   _____

Procedure

_____

_____

_____

_____

_____

_____

_____

_____

_____

_____

Observations and Results

Object	Amount of Force in ____

_____

_____

_____

_____

_____

_____

_____

_____

Conclusion

_____

_____

_____

_____

Written Assignment Week 1

Discussion Questions

Forces pp. 114-115

1. What does a force do? Name several examples.
2. Where is a force field the strongest?
3. What is the difference between contact and non-contact forces?

Combining Forces pg. 116

1. What is a resultant force?
2. How do you find the resultant when forces are pulling in the same direction?
3. How do you find the resultant when forces are pulling in the opposite direction?

Balanced Forces pg. 117

1. How is an object balanced?
2. Why is balance important to architects?

_____

_____

_____

_____

_____

_____

_____

_____

_____

_____

_____

_____

_____

_____

_____

_____

_____

_____

_____

Written Assignment Week 1

Student Assignment Sheet Week 2
Friction and Gravity

Experiment: How does friction affect movement?
Materials
- ✓ Force Meter from last week
- ✓ Small wooden block (aka. Jenga block)
- ✓ Eye-hook screw
- ✓ Sandpaper
- ✓ Felt
- ✓ Foil
- ✓ Spray oil
- ✓ Tape measure

Procedure
1. Read the introduction to the experiment and answer the question for the hypothesis section.
2. Screw the eye-hook screw into the top of the wooden block. Then, attach it to the hook on the force meter so that the block can be dragged horizontally. Next, use the tape measure to mark off a 1 foot (0.3 meter) track on a smooth surface, like a table our counter.
3. Now, place the block at the beginning of your track with the force meter in front over the track. Pull the block from the force meter evenly to the end in three seconds. Observe how much the rubber band on the force meter stretched and record that on your experiment sheet.
4. Then, place the piece of sandpaper on your track. Like before, put block at the beginning of the track and pull it evenly to the end in three seconds. Observe how much the rubber band on the force meter stretched and record that on your experiment sheet. Repeat with the felt.
5. Finally, place the foil over the track and coat it well with spray oil. Then, as before, put block at the beginning of your track and pull it evenly to the end in three seconds. Observe how much the rubber band on the force meter stretched and record that on your experiment sheet.
6. Draw conclusions and complete the experiment sheet.

Vocabulary & Memory Work
- ☐ Vocabulary: air resistance, friction, gravity, terminal velocity
- ☐ Memory Work—This week, begin working on memorizing Newton's three laws of motion. (*See Unit Overview Sheet for a complete listing.*)

Sketch: Types of Friction (*See the Sketch Notes.*)
- ▨ Label the following – Static friction, sliding friction, rolling friction, fluid friction

Writing
- ᐕ Reading Assignment: *DK Encyclopedia of Science* pg. 121 Friction, pg. 122 Gravity
- ᐕ Additional Research Readings
  - 📖 Relativity and Gravity: *KSE* pp. 298-299
  - 📖 Friction: *KSE* pp. 308-309
  - 📖 Gravitation: *UDIS* pp. 18-19

Dates
- ☉ 1630's – Galileo does a series of experiments with a marble and a series of differently-shaped tracks, which leads to the discovery of a retarding force called friction.
- ☉ 1955 – Christopher Cockerell invents the hovercraft, which uses a cushion of air that allows a vehicle to move without friction.

Sketch Week 2

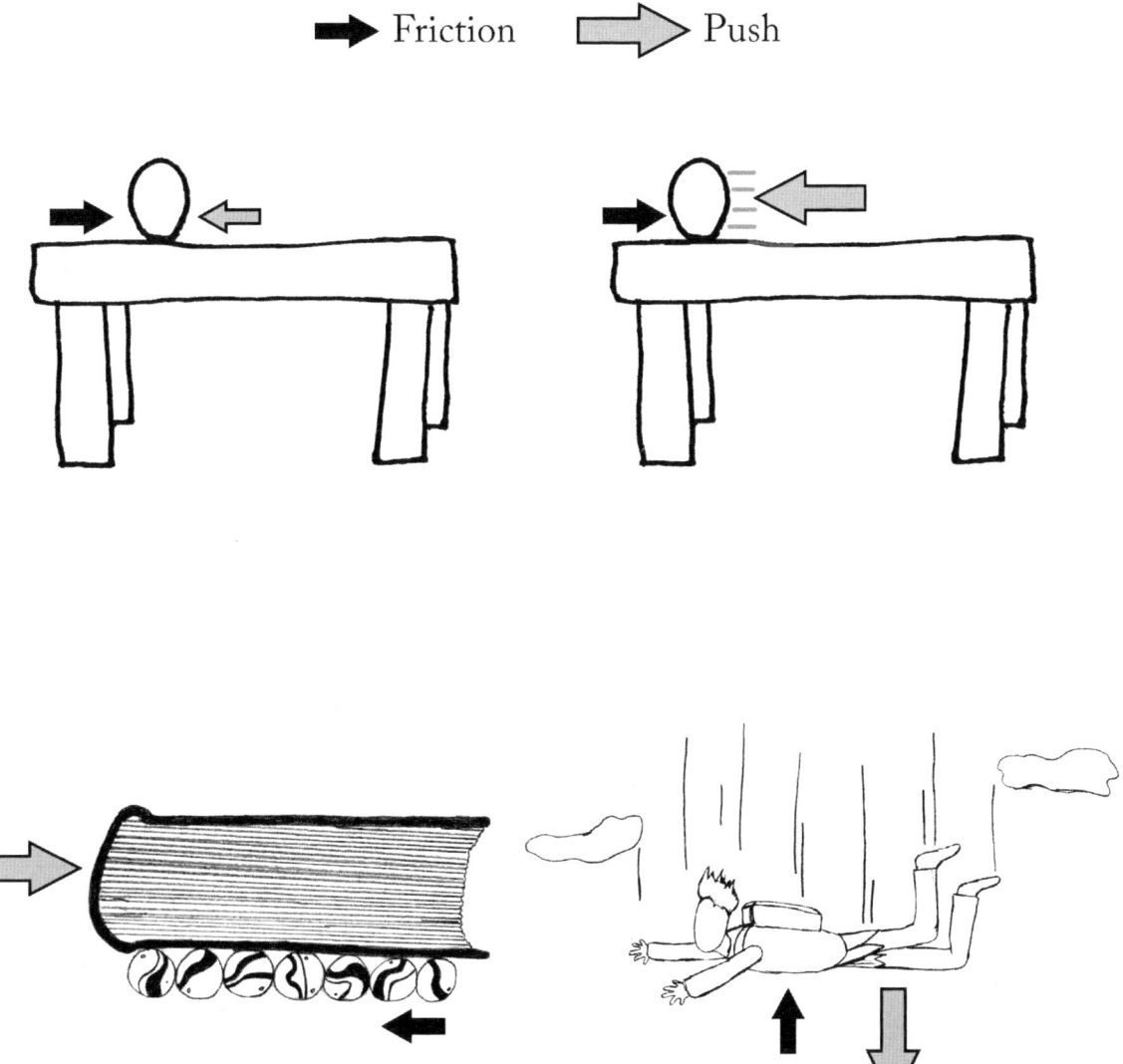

How does friction affect movement?

Introduction

When an object is in forward motion, several forces are acting on it. There is the driving force, which is propelling the object forward. There is weight (or gravity), which pulls the object downward. There is air resistance, which slows the object down. Finally, there is friction. In today's experiment, you are going to act as the driving force for a block as it moves across a track. Then, you are going to use a variety of materials to test how friction affects the motion of the block.

Hypothesis

_____

_____

_____

Materials

_____   _____

_____   _____

_____   _____

Procedure

_____

_____

_____

_____

_____

_____

_____

_____

_____

_____

_____

_____

Observations and Results

	On Smooth Surface	On Sandpaper	On Felt	On Foil With Oil
Amount of Force in ___				

_____

_____

_____

_____

Conclusion

_____

_____

_____

_____

_____

_____

_____

Written Assignment Week 2

Discussion Questions

Friction, pg. 121

    1. How does the roughness of a surface affect the amount of friction?

    2. Why is friction so important?

    3. What does a streamlined design do?

Gravity, pg. 122

    1. What two things affect the force of gravity?

    2. What is the center of gravity?

    3. How does gravity cause the tides in the ocean?

Written Assignment Week 2

Student Assignment Sheet Week 3
Motion

Experiment: Investigating the Three Laws
Materials
- ✓ Jenga block with the eyehook from last week
- ✓ String
- ✓ 2 Toy cars
- ✓ Egg

Procedure
1. In this experiment, you will be investigating the three laws of motion. Begin by reading the introduction.
2. Test the three laws of motion.
   - **Motion Law # 1** – You will need a Jenga block with the eyehook and string. Tie the string to the block and place it on a smooth surface. Pull the block along with a decent amount of force and then stop suddenly. Observe what happens to the block.
   - **Motion Law # 2** – You will need the two toy cars and a partner for this test. Line up the two cars evenly on a flat surface. Gently push your car forward, while your partner pushes his car forward with a greater force at the same time. Observe what happens to the two cars.
   - **Motion Law # 3** – You will need an egg for this test. Head outside with the egg. Drop the egg onto the pavement from a height of four to five feet and observe what happens.
3. Draw conclusions and complete the experiment sheet.

Vocabulary & Memory Work
- ☐ Vocabulary: inertia, mass, weight, momentum
- ☐ Memory Work — This week, continue working on memorizing Newton's three laws of motion. Also work on memorizing the following equation from Newton's second law:
  - ↳ Force (F) = mass (m) • acceleration (a)

Sketch: 3 Laws of Motion
- ▨ Label each of the three sketches with the law of motion that they represent. (*See the experiment sheet for a list of the laws.*)

Writing
- ᕲ Reading Assignment: *DK Encyclopedia of Science* pg. 120 Forces and Motion
- ᕲ Additional Research Readings
  - 📖 Momentum: *KSE* pp. 296-297
  - 📖 Dynamics: *UDIS* pp. 12-13

Dates
- ☉ 1665 – The plague breaks out in London, which forces Isaac Newton to leave Trinity College in Cambridge. He goes home and spends the next two years working on his book, *Principia*, in which he shares his three laws of motion.

Sketch Week 3

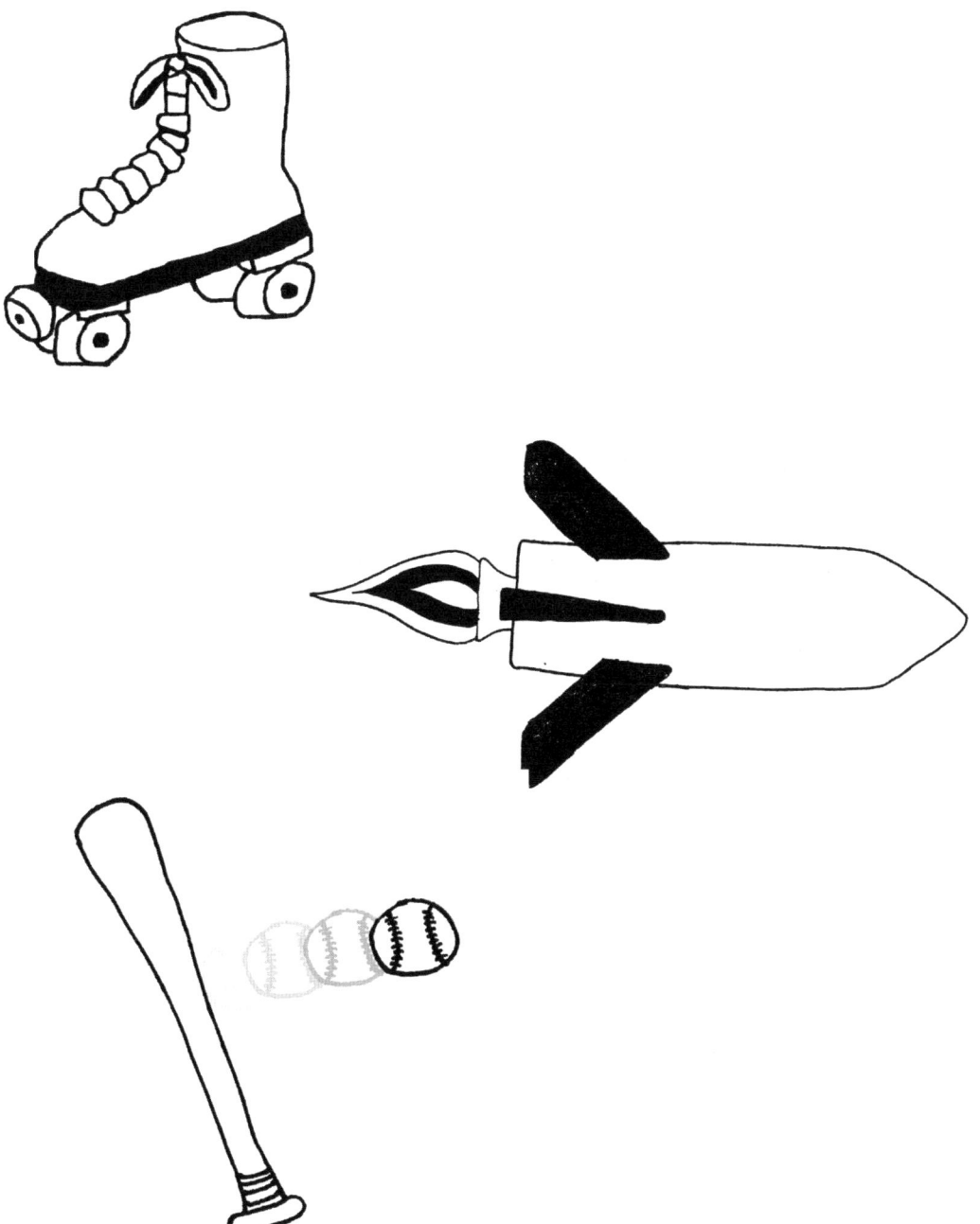

Investigating the Three Laws

Introduction

Isaac Newton built on Galileo's work on friction and motion through number of experiments. These tests led to his development of the three laws of motion. The laws state:

1. An object will not move, unless a force like a push or pull moves it. Once it is moving, an object will not stop moving in a straight line unless it's forced to change.
2. The greater the force on an object, the greater the change in its motion. The greater the mass of an object, the greater the force needed to change its motion.
3. For every reaction, there is an equal but opposite reaction.

In today's experiment, you are going to do three tests where you will see each of the laws in action.

Motion Test #1

Procedure

_____

_____

_____

_____

_____

Observation

_____

_____

_____

_____

_____

Motion Test #2

Procedure

_____

_____

_____

_____

_____

Observation

_____

_____

_____

_____

_____

Motion Test #3

Procedure

_____

_____

_____

_____

_____

Observation

_____

_____

_____

_____

_____

Conclusion

_____

_____

_____

_____

_____

_____

Written Assignment Week 3

Discussion Questions

1. What did Aristotle believe about motion?
2. What did Galileo learn about motion?
3. What did Isaac Newton discover about motion?

Written Assignment Week 3

Student Assignment Sheet Week 4
Speed and Acceleration

Experiment: Will the height of the ramp affect a car's speed?

Materials

- ✓ Cardboard or plastic track
- ✓ Blocks or thick books
- ✓ Toy car
- ✓ Stopwatch

Procedure

1. Read the introduction to the experiment and answer the question in the hypothesis section.
2. Use the cardboard (or plastic track) to build a track that is 1 meter long. Use the blocks (or books) to prop the track up so that one end of it is 15 centimeters higher than the other. Now, use the stopwatch to measure the time it takes for the car to go from the top of the track to the end. Record the time on the experiment sheet and then repeat two more times for a total of three trials at 15 centimeters.
3. Next, add several more blocks (or books) to prop the track up so that one end of it is 30 centimeters higher than the other. Use the stopwatch to measure the time it takes for the car to go from the top of the track to the end. Record the time on the experiment sheet and then repeat two more times for a total of three trials at 30 centimeters.
4. Finally, add few more blocks (or books) to prop the track up so that one end of it is 45 centimeters higher than the other. Use the stopwatch to measure the time it takes for the car to go from the top of the track to the end. Record the time on the experiment sheet and then repeat two more times for a total of three trials at 45 centimeters.
5. Draw conclusions and complete the experiment sheet.

Vocabulary & Memory Work

- ☐ Vocabulary: acceleration, speed, velocity
- ☐ Memory Work—This week, continue working on memorizing Newton's three laws of motion. Also work on memorizing the following equations for velocity and acceleration:
  - ↳ Speed (v) = total distance (d) / total time (t)
  - ↳ Acceleration (a) = change in velocity ($v_f - v_i$) /total time (t)

Sketch: Acceleration Graph

- Label the following parts of the line on the graph – constant acceleration, constant speed, and constant deceleration

Writing

- ✑ Reading Assignment: *DK Encyclopedia of Science* pg. 118 Speed, pg. 119 Acceleration
- ✑ Additional Research Readings
  - 📖 Motion: *UDIS* pp. 10-11

Dates

- 🕐 1905 – Albert Einstein publishes his theory of relativity, which is the basis for many of the ideas we have about our universe.

Sketch Week 4

Acceleration Graph

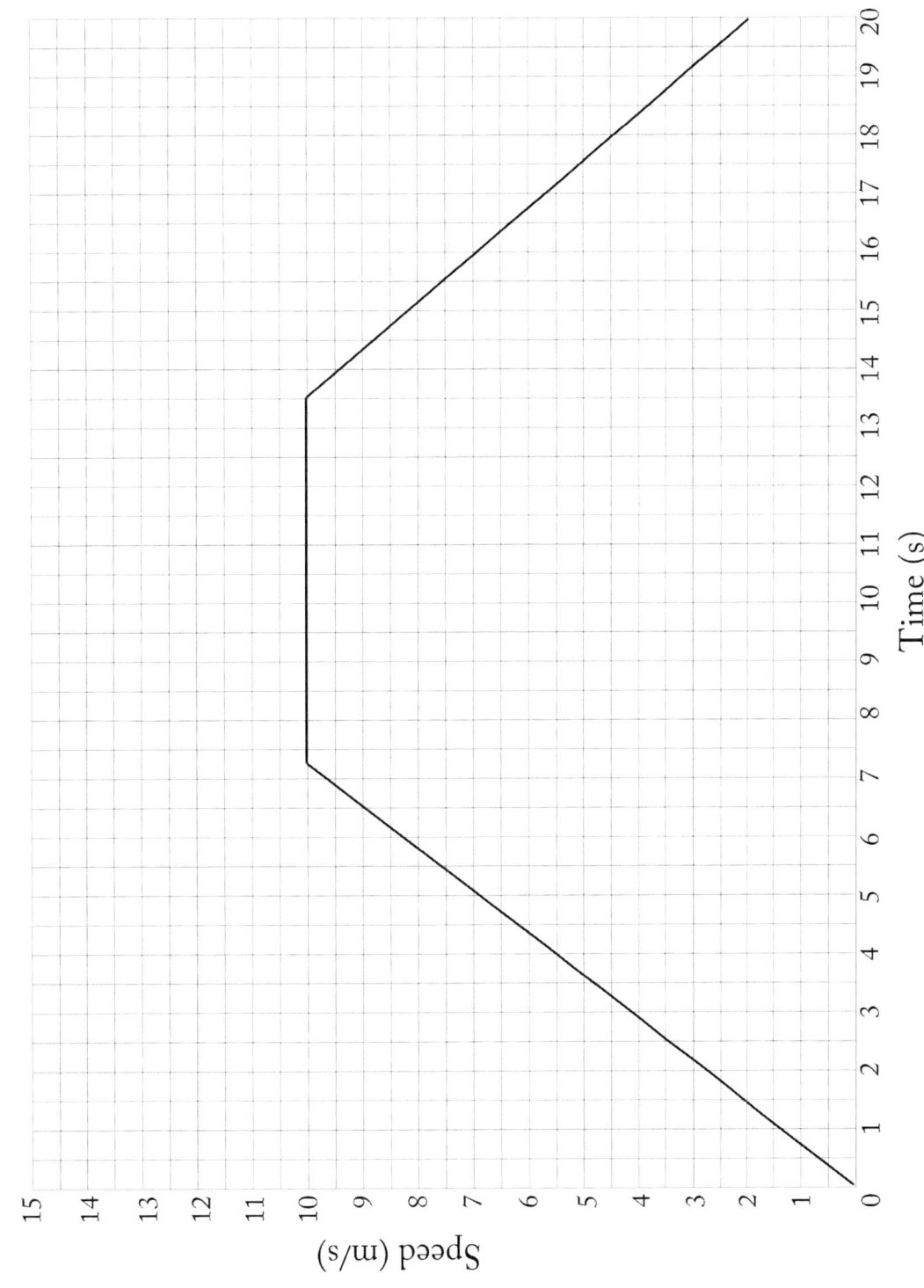

Speed (m/s) — Time (s)

Will the height of the ramp affect a car's speed?

Introduction

We know from Newton's laws of motion that a force must act on a vehicle to get it moving. We also know that the greater the force, the quicker the vehicle will go. In today's experiment, you are going to measure the time it takes for a toy car to go down a given length of track. You will also vary the height of the track to see how the different angles increase or decrease force of gravity affecting the speed of the car.

Hypothesis

_____

_____

_____

Materials

_____   _____

_____   _____

_____   _____

_____   _____

_____

Procedure

_____

_____

_____

_____

_____

_____

_____

_____

_____

_____

_____

Observations and Results

Track Height	Trial #1	Trial #2	Trial #3	Average Time
Track at Height of 15 cm				
Track at Height of 30 cm				
Track at Height of 45 cm				

_____

_____

_____

Conclusion

_____

_____

_____

_____

_____

_____

_____

_____

Written Assignment Week 4

Discussion Questions

Speed, pg. 118
1. What is the average speed?
2. What is instantaneous speed?
3. How are speed and velocity related?

Acceleration, pg. 119
1. What is acceleration?
2. What is deceleration?

_____

_____

_____

_____

_____

_____

_____

_____

_____

_____

_____

_____

_____

_____

_____

_____

_____

_____

_____

Written Assignment Week 4

Physics
Unit 2

Energy

Unit 2 Energy
Vocabulary Sheet

Define the following terms as they are assigned on your Student Assignment Sheet.

1. Energy – _____

_____

_____

2. Kinetic energy – _____

_____

_____

3. Potential energy – _____

_____

_____

4. Work – _____

_____

_____

5. Fluid – _____

_____

_____

6. Pressure – _____

_____

_____

7. Non-renewable energy resources – _____

_____

_____

8. Renewable energy resources – _____

_____

_____

9. Fulcrum – _____

_____

_____

10. Input force – _____

_____

_____

11. Output force – _____

_____

_____

Student Assignment Sheet Week 5
Energy

Experiment: Do different types of food contain different amounts of energy?
Materials

- ✓ Goldfish cracker
- ✓ Small marshmallow
- ✓ Piece of lettuce
- ✓ Piece of bacon fat
- ✓ Aluminum pan
- ✓ Matches
- ✓ Safety glasses
- ✓ Bucket of water

Procedure

Caution – Be sure to do this experiment in a well-ventilated area with a fire extinguisher close at hand.

1. Read the introduction to the experiment and answer the question for the hypothesis section.
2. Head outside and set the aluminum pan on a concrete or tile surface or inside a grill. Place the cracker in the aluminum pan and have an adult use the match to light the cracker. Time how long it takes for the cracker to burn and write this on your experiment sheet.
3. Use the bucket of water to extinguish any smoldering embers. Then, clean and dry the aluminum pan.
4. Now, repeat the procedure from step 2 and 3 for the marshmallow, lettuce, and bacon fat. (**Note**—*Make sure that your samples are similar in size, so that your results are more accurate.*)
5. Draw conclusions and complete the experiment sheet.

Vocabulary & Memory Work

- ☐ Vocabulary: energy, potential energy, kinetic energy, work
- ☐ Memory Work—Begin working on memorizing the types of energy (see Appendix pg. __) along with the following equation:
  - ↳ Work (W) = Force (F) • Distance (d)

Sketch: Energy Diagram

- ▨ Label the following – Energy, kinetic energy, energy from motion; potential energy, energy from gravity, energy from a stretched or compressed object, energy from an electrical charge, energy from the pull of a magnet

Writing

- ꝏ Reading Assignment: *DK Encyclopedia of Science* pp. 132-133 Work and Energy
- ꝏ Additional Research Readings
  - 📖 Potential and Kinetic Energy: *KSE* pp. 292-293
  - 📖 Energy: *UIDS* pp. 8-9

Dates

- ☉ 1818-1889 – James Joule lives. He is credited with discovering that work produces heat, which is a form of energy.

Sketch Week 5

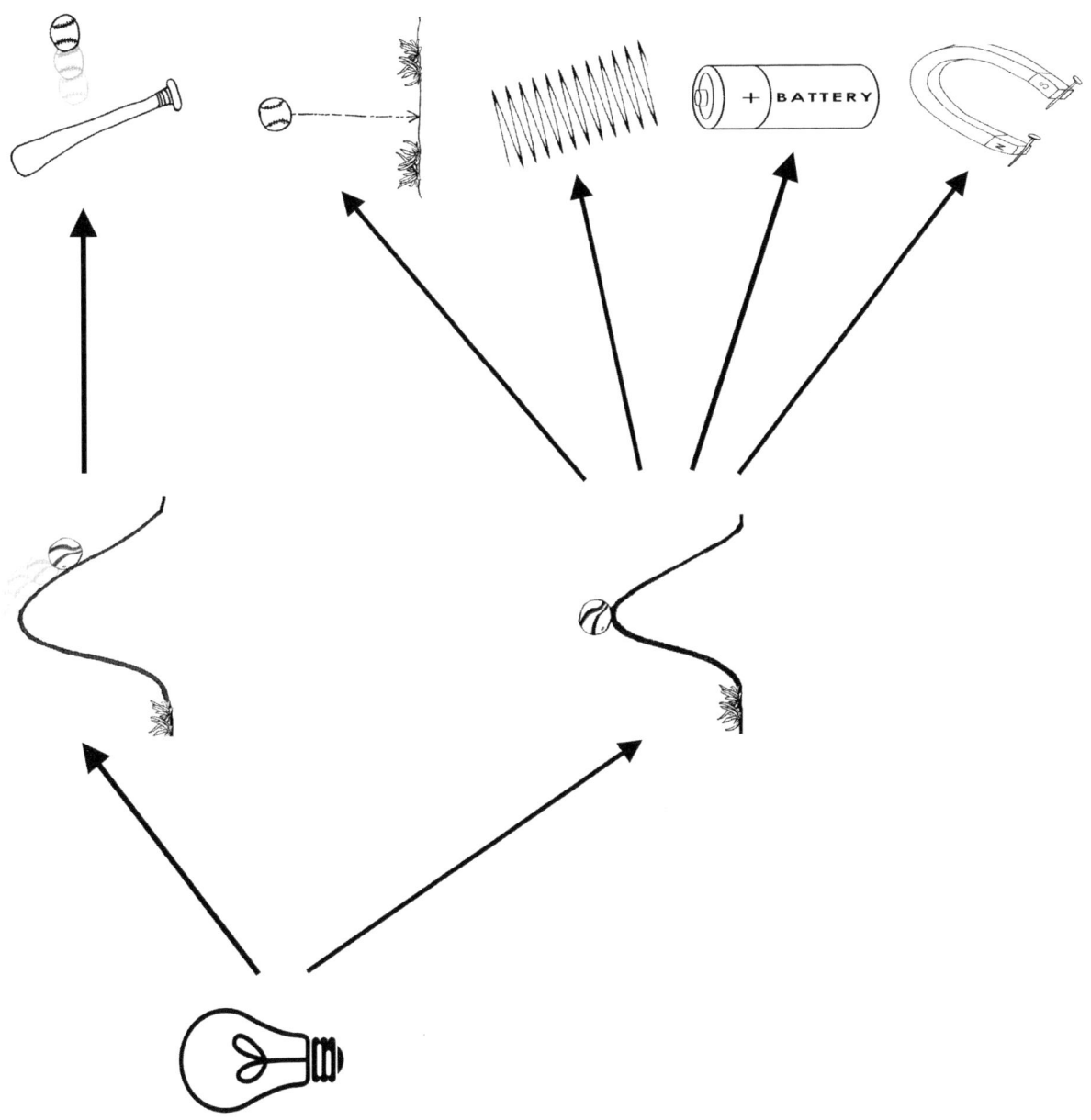

Do different types of food contain different amounts of energy?

Introduction

As we break down the food we eat, it releases energy in the form of calories. Some of the chemicals in food are broken down quickly, releasing a burst of energy. Other chemicals take a bit more work to break down, so they release energy slowly over time. These reactions work together to fuel our bodies' processes over time. In today's experiment, you will be testing different types of food to see how much energy they release.

Hypothesis

_____

_____

_____

Materials

_____   _____

_____   _____

_____   _____

_____   _____

_____   _____

Procedure

_____

_____

_____

_____

_____

_____

_____

_____

_____

Observations and Results

Type of Food	Burn Time

_____
_____
_____
_____
_____
_____
_____
_____

Conclusion

_____
_____
_____
_____

Written Assignment Week 5

Discussion Questions

1. What happens when work is done?
2. How are work and energy related?
3. What are some common forms of energy?
4. What are the two units of measurement for energy?
5. What is the difference between kinetic and potential energy?

Written Assignment Week 5

Student Assignment Sheet Week 6
Pressure

Experiment: What affects the pressure exerted by a fluid?

Materials

- ✓ 2-Liter Soda bottle
- ✓ 2 Cans – one large, one small
- ✓ Screw
- ✓ Water
- ✓ Piece of clay
- ✓ Cup measure
- ✓ Tape measure

Procedure

1. Read the introduction to the experiment and answer the question for the hypothesis section.
2. Perform the two pressure tests.
   - **Pressure Test #1** – Fill the small tin can with water and then use a cup measure to determine how much water the can holds. Use the screw to poke a hole in the bottom of the small tin can about ½ inch (1.25 cm) from the bottom. Put a piece of clay over the hole to prevent any water spilling out before filling the can with water. Lay out the tape measure on a surface that you don't mind getting wet, preferably outdoors. Place the can at the first mark on the tape measure. Quickly remove the clay plug. Measure how far the water goes and record it on the experiment sheet. Repeat this process two more times and then average the distances. Repeat steps 2 and 3 using the larger can.
   - **Pressure Test #2** – Use the screw to poke three holes in the 2-Liter bottle – one about 1 inch (2.5 cm) from the bottom, one at the middle of the bottle, and one 2 inches (5 cm) from the top of the bottle. Plug each of the holes with clay and the fill the bottle to the very top. Place the full 2-Liter bottle at the first mark on the tape measure. Quickly remove the clay plugs. Measure how far the water goes from each of the holes and record it on the experiment sheet. Repeat this process two more times and then average the distances.
3. Draw conclusions and complete the experiment sheet.

Vocabulary & Memory Work

- ☐ Vocabulary: fluid, pressure
- ☐ Memory Work—Continue to work on memorizing the different types of energy along with the following equation:
  - ↯ Pressure (P) = $\dfrac{\text{Force (F)}}{\text{Area (a)}}$

Sketch: Pressure Relationships

- ▨ Label the following – Pressure and force relationship, increase force, increase pressure, decrease force, decrease pressure; pressure and area relationship, decrease area, increase pressure, increase area, decrease pressure
- ▨ Fill in the equation at the bottom of the sketch.

Writing

- ✑ Reading Assignment: *DK Encyclopedia of Science* pg. 127 Pressure
- ✑ Additional Research Readings
  - 📖 Pressure: *KSE* pg. 311, *UIDS* pg. 25; Fluids: *KSE* pg. 310

Dates

- ☉ 1643 – Evangelista Torricelli invents the mercury barometer.

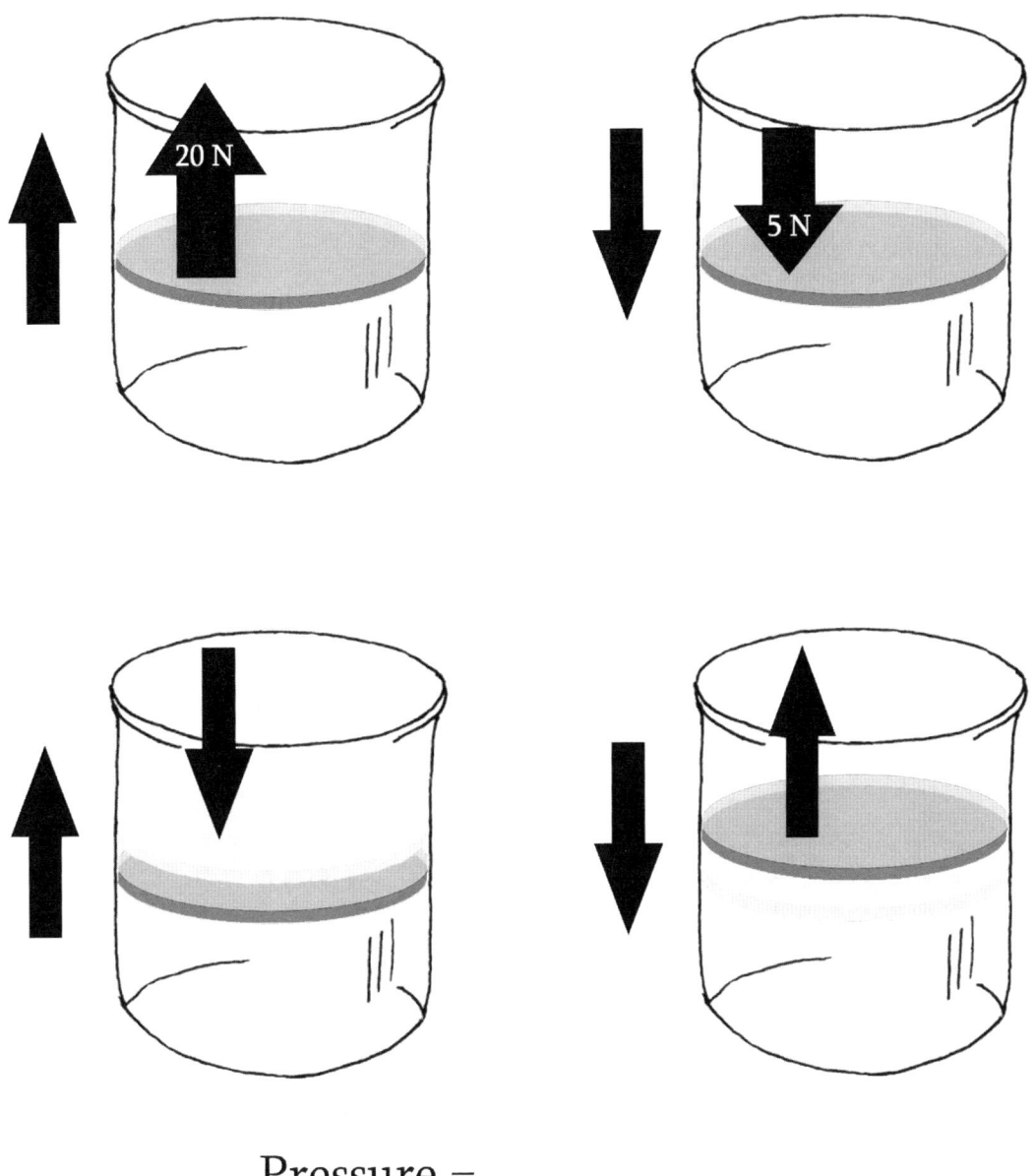

Pressure = _____

What affects the pressure exerted by a fluid?

Introduction

A fluid, such as a liquid or a gas, fills in the available space within a container. The fluid exerts pressure on the surfaces of the container throughout the space it occupies. The amount of that pressure depends upon several factors. In today's experiment, you will be looking at two factors – area and force – to see how they affect the pressure a fluid exerts in a container.

Hypothesis

_____

_____

_____

Materials

_____   _____

_____   _____

_____   _____

Pressure Test #1

Procedure

_____

_____

_____

_____

_____

Observations and Results

	Trial #1	Trial #2	Trial #3	Average Distance
Small Can				
Large Can				

_____

_____

_____
_____
_____

Pressure Test #2

Procedure

_____
_____
_____
_____
_____

Observations and Results

	Trial #1	Trial #2	Trial #3	Average Distance
Top Hole				
Middle Hole				
Bottom Hole				

_____
_____
_____
_____
_____

Conclusion

_____
_____
_____
_____
_____
_____

Written Assignment Week 6

Discussion Questions

1. How are pressure, force, and area related?
2. What happens to air pressure as you go high up? Why?
3. What happens to water pressure as you dive deep under water? Why?
4. Why do we not feel the pressure of the surrounding air on our bodies?

_____

_____

_____

_____

_____

_____

_____

_____

_____

_____

_____

_____

_____

_____

_____

_____

_____

_____

_____

_____

Written Assignment Week 6

_____

Student Assignment Sheet Week 7
Energy Sources

Experiment: Build a Solar Oven

    Materials
- ✓ Foil
- ✓ Black construction paper
- ✓ Small cardboard box (about 9" by 12")
- ✓ Plastic wrap
- ✓ Tape
- ✓ Marshmallow
- ✓ Small glass dish (one that will fit inside the box)

    Procedure
1. Read the introduction to the experiment.
2. Place the black construction paper on the bottom of your box. This will serve to absorb the sunlight.
3. Tape the flaps down. Then, cover the sides with the foil so that the shiny side is facing out. The foil will help to reflect the sunlight and focus it on cooking the marshmallow.
4. Put the glass dish in the center of the bottom of the box and set the marshmallow on it. Cover the top box with plastic wrap; this will help to trap the heat.
5. Now, head outside and set your solar oven in direct sunlight. Observe what happens.
6. Draw conclusions and complete the experiment sheet.

Vocabulary & Memory Work
- ☐ Vocabulary: non-renewable energy resources, renewable energy resources
- ☐ Memory Work—Continue to work on memorizing the different types of energy.

Sketch: Energy Sources Chart
- Create a chart depicting renewable and non-renewable energy sources.

Writing
- Reading Assignment: *DK Encyclopedia of Science* pp. 134-135 Energy Sources
- Additional Research Readings
  - Harnessing Wave and Wind Power: *KSE* pp. 328-331

Dates
- 🕐 c. 100 – The Romans began to burn coal as a source of energy.
- 🕐 c. 650 – The Persians began to use windmills as a source of energy.
- 🕐 1891 – Hydroelectric power is first demonstrated in Germany.
- 🕐 1960 – Turkmenistan builds the first solar thermal power plant.

Sketch Week 7

Renewable

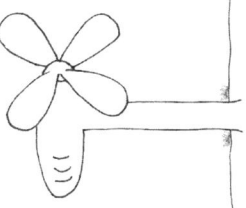

Energy Sources

Non-renewable

Build a Solar Oven

Introduction

On Earth, we have many different sources of energy. Some of these sources are not renewable, such as fossil fuels like oil and coal. Some of the sources are renewable, like wind, waves, and the sun. In this experiment, you are going to build a device that will allow you to harness the power of the sun to cook a marshmallow.

Materials

_____   _____

_____   _____

_____   _____

Procedure

_____

_____

_____

_____

_____

_____

_____

_____

_____

_____

_____

_____

Observations and Results

_____

_____

_____

_____

My Solar Oven

Conclusion

_____

_____

_____

_____

_____

_____

_____

Written Assignment Week 7

Discussion Questions

1. What is the main source of renewable energy on the earth?
2. What is the difference between renewable and non-renewable energy sources?
3. How do solar panels work?
4. How can wind and water create power?
5. How does a power plant work?
6. What are fossil fuels?

Written Assignment Week 7

Student Assignment Sheet Week 9
Simple Machines

Experiment: Build a Simple Machine

Materials

✓ Materials will vary based on what you choose to build.

Procedure

1. Read the introduction to the experiment.
2. Choose one of the types of simple machines (lever, wheel and axle, gear, inclined plane, wedge, screw, or pulley) to build. (**Note**—*See the experiment sheet for an explanation of each of the types of simple machines.*)
3. Construct the machine and then test to see if it helps to perform work. For instance, if you choose to build a lever, test it by using a stack of books. First, pick up the stack on your own, then use the lever to pick up the stack. Which was easier to do?
4. Draw conclusions and complete the experiment sheet.

Vocabulary & Memory Work

- ☐ Vocabulary: fulcrum, input force, output force
- ☐ Memory Work—Continue to work on memorizing the different types of energy.

Sketch: Types of Simple Machines

- ▣ Label the following – lever, wheel and axle, inclined plane, wedge, screw, pulley, gear

Writing

- ⌇ Reading Assignment: *DK Encyclopedia of Science* pp. 130-131 (Machines)
- ⌇ Additional Research Readings
  - 📖 Ramps and Wedges: *KSE* pg. 300
  - 📖 Levers and Pulleys: *KSE* pg. 301
  - 📖 Wheels and Axles: *KSE* pg. 302
  - 📖 Machines: *UIDS* pp. 20-21

Dates

- ⊕ c. 3rd century BC – Archimedes is said to have invented a screw pump to help get water from a reservoir source to the fields for irrigation.

Sketch Week 9

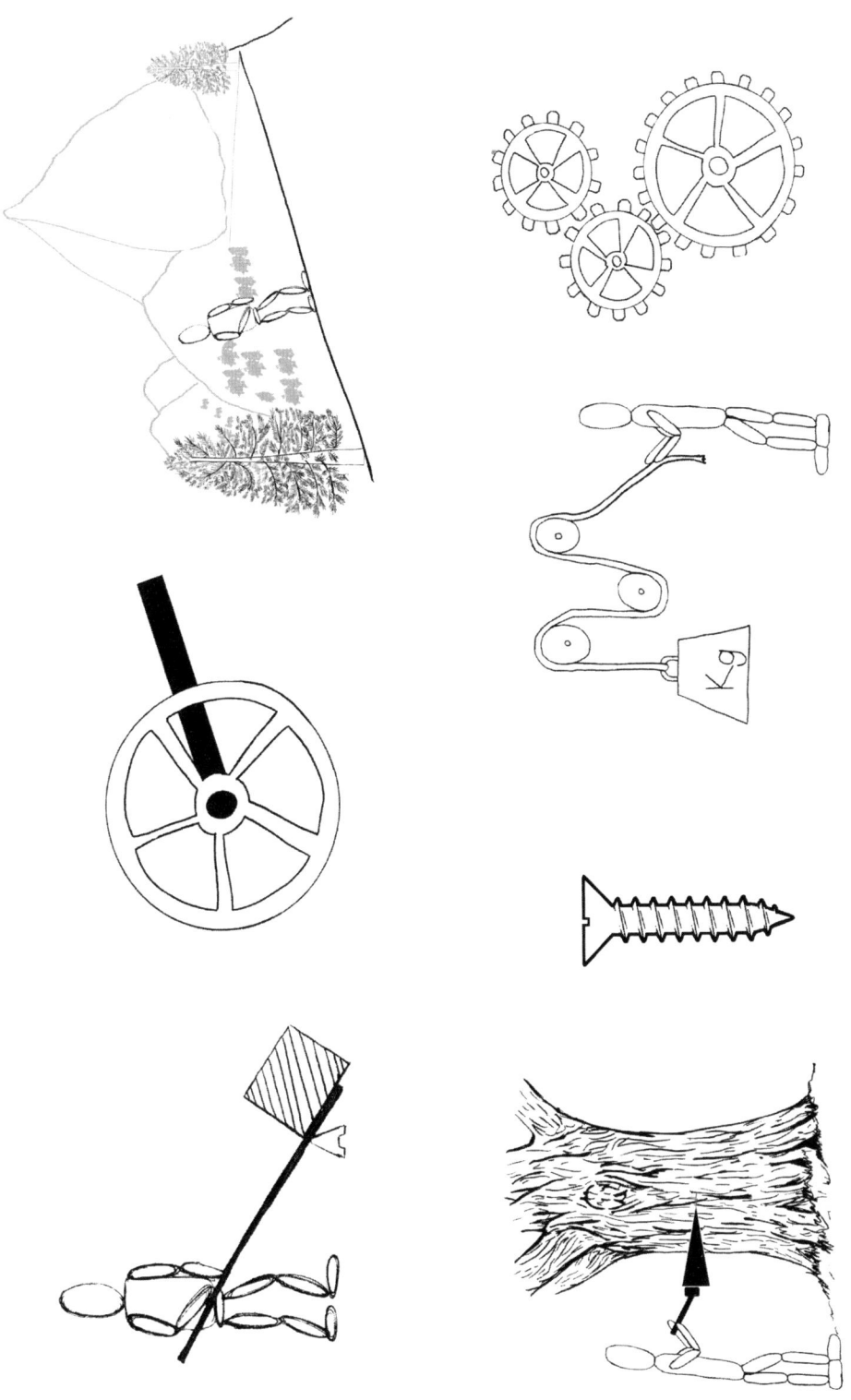

Build a Simple Machine

Introduction

A simple machine is a tool that you can use to help you to do work. In other words, simple machines make the task of lifting or moving an object easier. There are seven main types of simple machines.

- **Lever** – A rigid bar that is free to move around at a fixed point. Levers are frequently used to lift things, i.e. using a flat piece of metal to pry open a paint can, using a wheelbarrow to move dirt, or using a shovel to life a large rock.
- **Wheel and axle** – Two disks or cylinders, each with a different radius. The wheel and axle simple machine is often used to turn or rotate things, i.e. using a screwdriver to turn a screw or using the steering wheel of a vehicle.
- **Gears** – Toothed wheels that interlock in pairs; each one helps to drive the next. Gears are often used to consistently drive an object, like a watch or clock.
- **Inclined plane** – A slanted surface that helps move objects up an incline. The inclined planes are typically ramps that can be used to move a heavy object from a lower level to a higher level.
- **Wedge** – A v-shaped object whose sides are two inclined planes. Wedges, like knives, axes, and zippers, make it easier to separate two objects.
- **Screw** – An inclined plane wrapped around a cylinder; this plane is also called the thread of the screw. These simple machines, including screws, nuts, and bolts, make it easier to drive an object into another object.
- **Pulley** – A rope that fits in the groove of a wheel. Pulleys make it easier to pull a heavy load directly upward.

In today's experiment, you are going to choose one of these types of simple machines to build.

Materials

Procedure

_____

_____

_____

Observations and Results

```
+-----------------------------------------------+
|              My Simple Machine                |
|                                               |
|                                               |
|                                               |
|                                               |
|                                               |
|                                               |
|                                               |
|                                               |
|                                               |
|                                               |
+-----------------------------------------------+
```

_____

_____

_____

Conclusion

_____

_____

_____

_____

_____

Written Assignment Week 9

Discussion Questions

1. What can machines do?
2. What is a complex machine?
3. What is the main benefit of using simple machines?
4. Name several examples of simple machines.

Written Assignment Week 9

Physics
Unit 3

Thermodynamics

Unit 3 Thermodynamics
Vocabulary Sheet

Define the following terms as they are assigned on the Student Assignment Sheet.

1. Energy conversion – _____

   _____

   _____

2. Entropy – _____

   _____

   _____

3. Heat – _____

   _____

   _____

4. Temperature – _____

   _____

   _____

5. Absolute zero – _____

   _____

   _____

6. Conduction – _____

   _____

   _____

7. Convection – _____

   _____

   _____

8. Radiation – _____

   _____

   _____

9. Internal combustion engine – _____

_____

_____

10. Power – _____

_____

_____

11. Steam engine – _____

_____

_____

Student Assignment Sheet Week 9
Energy Conversion

Science Fair Project

This week, you will complete step one and begin step two of your Science Fair Project. You will be choosing your topic, formulating a question and doing some research about that topic.

1. **Choose your topic** – You should choose a topic in the field of physics that interests you, such as rockets. Next, come up with several questions you have relating to that topic, (e.g., "How does the design of the rocket affect its flight?" or "What type of rocket flies the highest?"). Then, choose the one question you would like to answer and refine it (e.g., "How does the fin design affect the rocket's flight?").

2. **Do Some Research** – Now that you have a topic and a question for your project, it is time to learn more about your topic so that you can make an educated guess (hypothesis) about the answer to your question. For the question stated above, you would need to research topics like rockets, fin designs, and flight. Begin by looking up the topic in the references you have at home. Then, make a trip to the library to search for more on the topic. As you do your research, write any relevant facts you have learned on index cards and be sure to record the sources you use.

Vocabulary & Memory Work

☐ Vocabulary: energy conversion, entropy

☐ Memory Work—This week, begin working on memorizing the laws of thermodynamics. (*See Appendix pg. __ for a complete listing.*)

Sketch: Energy Chain

▨ Label the following – sun, nuclear energy is changed into heat and light energy, plants, light is changed into chemical energy, animals, chemical energy is changed into kinetic energy used of activities like breathing and moving

Writing

✑ Reading Assignment: *DK Encyclopedia of Science* pp. 138-139 (Energy Conversion)

✑ Additional Research Reading

📖 Thermodynamics: *KSE* pg. 258 (Introduction and section on the First Law)

Dates to Enter

🕐 c.550 BC – Ancient philosopher, Thales Miletus, believes that there is a conservation of some sort of hidden substance of which everything is made. (**Note**—*Today we call this hidden substance mass energy.*)

🕐 1840's – James Joule does a number of experiments that lead to the development of the law of conservation of energy.

🕐 1850 – German scientist, Rudolf Clausius, suggests that the law of conservation of energy should be called the first law of thermodynamics.

🕐 1865 – Rudolf Clausius coins the term "entropy," which refers to unusable energy.

🕐 1920 – Ralph Fowler develops the zeroth law of thermodynamics.

Sketch Assignment Week 9

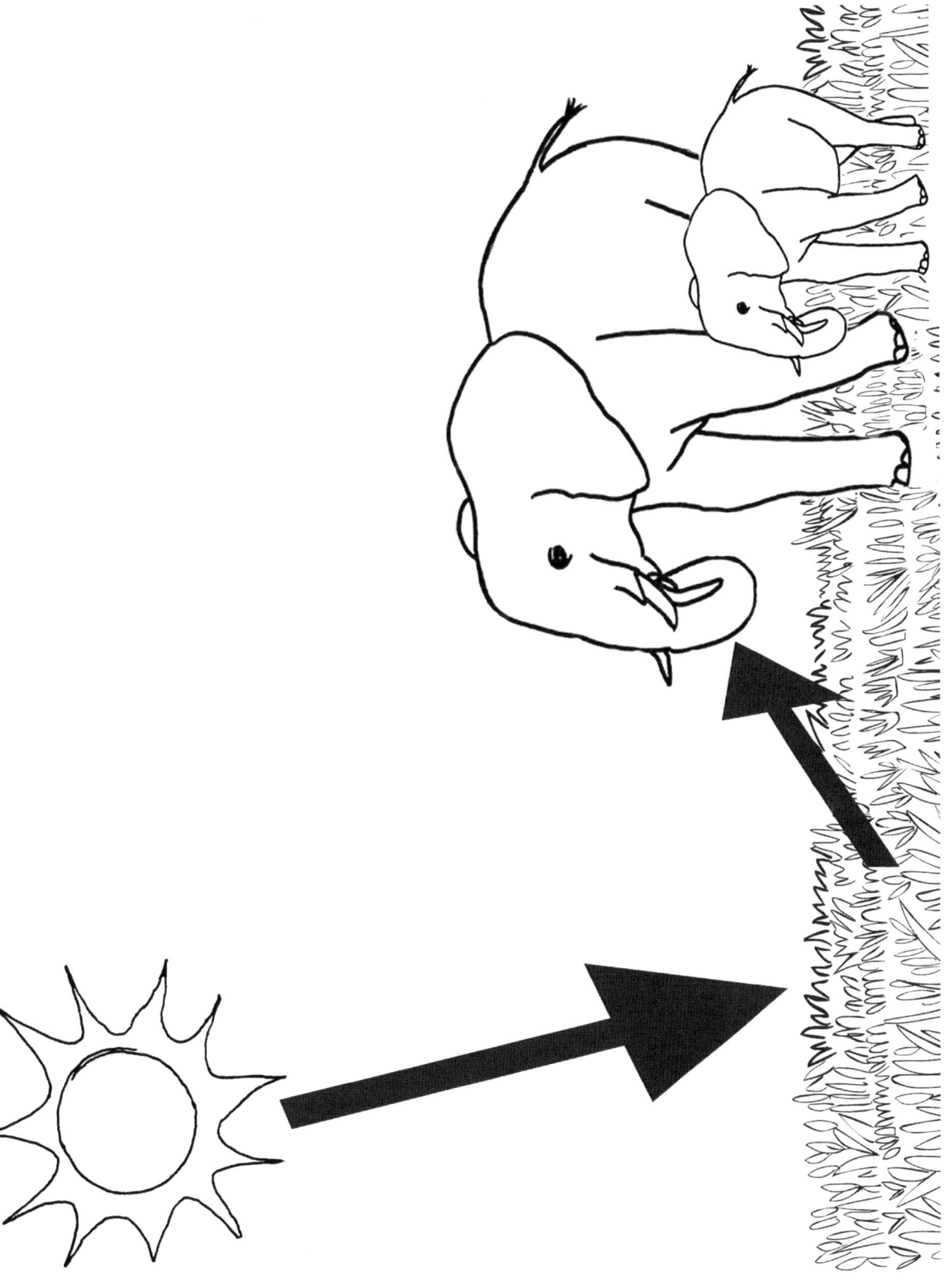

Science Fair Project Step 1: Choose a Topic

Key 1 ~ Decide on an area of science.

What areas of physics are you interested in learning about?

_____

_____

_____

_____

Rank your interest in the different areas you listed and then circle the one area that you would like to use for your topic.

Key 2 ~ Develop several questions about the area of physics.

What questions would you like to answer about your area of physics? (***Note***—*Remember that good questions begin with how, what, when, who, which, why, or where.*)

_____

_____

_____

_____

_____

_____

Key 3 ~ Choose a question to be the topic.

Write down the question that you will be using for your project.

_____

_____

_____

Science Fair Project Step 2: Do Some Research

Key 1 ~ Brainstorm for research categories.

What categories are you going to research for your project?

1. _____

2. _____

3. _____

4. _____

5. _____

Key 2 ~ Research the categories.

Use the following template for your research cards:

```
Category Number                    Reference Letter
_____
_____
           One piece of Information
_____
_____
_____
_____
_____
```

Record your sources below.

A. _____

_____

B. _____

_____

C. _____

_____

D. _____

_____

E. _____

_____

F. _____

_____

Written Assignment Week 9

Discussion Questions

1. Name one example of energy changing from one form to another.
2. What is an energy chain?
3. What always happens when energy is converted?
4. What is a perpetual motion machine? Why doesn't one exist?

_____

_____

_____

_____

_____

_____

_____

_____

_____

_____

_____

_____

_____

_____

_____

_____

_____

_____

_____

_____

_____

_____

_____

Written Assignment Week 9

Student Assignment Sheet Week 10
Heat

Science Fair Project

This week, you will complete steps two through four of your Science Fair Project. You will be finishing your research, formulating your hypothesis, and designing your experiment.

2. **Do Some Research** – This week, you will finish your research. Then, organize your research index cards and write a brief report on what you have found out.

3. **Formulate a Hypothesis** – A hypothesis is an educated guess. For this step, you need to review your research and make an educated guess about the answer to your question. A hypothesis for the question asked in step one would be, "The rounder and smaller the fins, the faster the rocket will go."

4. **Design an Experiment** – Your experiment will test the answer to your question. It needs to have a control and several test groups. Your control will have nothing changed, while your test groups will change only one factor at a time. An experiment to test the hypothesis given above would be make several different designs fins for your rocket – one that is rounded and small, one that is rounded and normal size, one that is rounded and large, one that is squared off and small, one that squared off and normal size, one that is squared off and large, plus the standard fins that come with the rocket. If time allows, you can go ahead and begin your experiment this week.

Vocabulary & Memory Work

☐ Vocabulary: heat, temperature, absolute zero

☐ Memory Work—This week, continue to work on memorizing the laws of thermodynamics and the Celsius to Fahrenheit equation ($°F = 1.8 \cdot °C + 32$).

Sketch

▨ There is no sketch this week, to allow for more time to work on the science fair project.

Writing

⌕ Reading Assignment: *DK Encyclopedia of Science* pp. 140-141 (Heat)

⌕ Additional Research Reading

📖 Temperature: *UIDS* pp. 26-27

📖 Thermodynamics: *KSE* pp. 258-259 (Read the remaining sections that you did not read last week.)

Dates to Enter

🕐 1714 – German physicist, Gabriel Fahrenheit, proposes a temperature scale where the lowest point was where he could cool brine and highest point was average internal temperature of the human body. It is eventually adopted and named after him.

🕐 1742 – Swedish physicist, Anders Celsius, develops a temperature scale where 0 represents the freezing point of water and 100 represents the boiling point of water. It is eventually adopted and named after him.

🕐 1848 – William Thomson, Lord Kelvin, develops an absolute temperature scale, known as the Kelvin scale.

Science Fair Project Step 2: Do Some Research

Key 3 ~ Organize the information.

Organize the information for your report.

Key 4 ~ Write a brief report.

Write down what the order of your categories will be for your report.

1. _____

2. _____

3. _____

4. _____

5. _____

Write out a rough draft of your research report on the following page.

Science Fair Project Step 3: Formulate a Hypothesis

Key 1 ~ Review the Research.

Read over your research.

Key 2 ~ Formulate an Answer.

Write down your hypothesis for your science fair project.

_____

_____

_____

_____

_____

Science Fair Project Step 4: Design an Experiment

Key 1 ~ Choose a Test.

What are some ways that you can test your hypothesis?

_____

_____

_____

_____

_____

_____

Key 2 ~ Determine the Variables.

What factor are we trying to test? (Independent variable)

_____

What factor will we use to measure the progress of our test? (Dependent variable)

_____

What factors do we need to keep constant so that they will not affect our results? (Controlled variables)

_____

_____

Key 3 ~ Plan the Experiment.

What will the groups in your experiment be?

Control Group: _____

Test Group 1: _____

Test Group 2: _____

Test Group 3: _____

Test Group 4: _____

Write down the plan for your experiment.

_____

_____

_____

_____

_____

_____

_____

Rough Draft of Science Fair Project Research Report

Written Assignment Week 10

1. What are the three temperature scales?
2. What is the difference between heat and temperature?
3. What happens to most objects as they are heated and cooled?
4. What is latent heat? And what happens to it as a material changes state?

Written Assignment Week 10

Student Assignment Sheet Week 11
Heat Transfer

Science Fair Project

This week, you will complete steps five and six of your Science Fair Project. You will carry out the experiment and record your observations and results.

5. **Perform the Experiment** – This week, you will perform the experiment you designed last week. Be sure to take pictures along the way as well as record your observations and results. (**Note**—*Observations are a record of the things you see happening in your experiment. For instance, an observation would be that the rocket with the wide, rounded fins wobbled quite a bit during flight. Results are specific and measurable. For instance, results would be that the rocket with wide, rounded fins flew 50 feet into the sky and landed 65 feet from the launch stand. Observations are generally recorded in journal form, while results can be compiled into tables, charts, and graphs or relayed in paragraph form.*)

6. **Analyze the Data** – Once you have compiled your observations and results, you can use them to answer your question. You need to look for trends in your data and make conclusions from that. A possible conclusion to the rocket-fin design experiment would be, "I found that the smaller and more rounded the fins, the further and smoother the rocket will fly." If your hypothesis does not match your conclusion or your were not able to answer your question using the results from your experiment, you may need to go back and do some additional experimentation.

Vocabulary & Memory Work

☐ Vocabulary: conduction, convection, radiation

☐ Memory Work—This week, continue to work on memorizing the laws of thermodynamics and begin working on the specific heat equation.

Sketch

▨ There is no sketch this week, to allow for more time to work on the science fair project.

Writing

✍ Reading Assignment: *DK Encyclopedia of Science* pg. 142 (Heat Transfer)

✍ Additional Research Reading

📖 Heat Transfer: *KSE* pp. 250-251

📖 Transfer of Heat: *UIDS* pp. 28-29

📖 Effects of Heat Transfer: *UIDS* pp. 30-31

Dates to Enter

🕐 1852 – William Thomson (Lord Kelvin) comes up with the idea of a "heat pump," a device that moves heat from a cold place to a hot one.

🕐 1892 – James Dewar invents the vacuum flask, which is designed to prevent the transfer of heat.

Science Fair Project Step 5: Perform the Experiment

Key 1 ~ Get ready for the experiment.

When do you plan to run your experiment?

From _____ to _____ .

Purchase and gather your materials, and prepare any of the materials that need to be pre-made.

Key 2 ~ Run the experiment.

What things do you need to remember to do each day?

Take pictures of the experiment every day or for every trial.

_____

_____

_____

_____

_____

_____

_____

_____

_____

Key 3 ~ Record any Observations and Results

Record your observations and results on the following page.

Science Fair Project Step 6: Analyze the Data

Key 1 ~ Review and organize the data.

What trends did you recognize in your observations?

_____

_____

_____

_____

_____

What information did you interpret from your results?

_____

_____

_____

_____

_____

_____

_____

Key 2 ~ State the answer.

After reviewing your data, write the answer to your question. (***Note****—Your statement should begin with "I found that…" or "I discovered that…"*)

_____

_____

_____

_____

_____

Key 3 ~ Draw several conclusions.

Answer the following questions:

✓ Was my hypothesis proven true? (**Note** — *If your hypothesis was proven false, be sure to state why you think it was proven false.*)
✓ Did you have any problems or difficulties when performing your experiment?
✓ Did anything interesting happen that you would like to share?
✓ Can you think of any other things related to your project that you would like to test in the future?

Now take your answer from key two and your answers from key three to write your conclusion on a separate sheet of paper. Your paragraph should be four to six sentences in length.

Science Fair Project Observations and Results

Observations

_____

_____

_____

_____

_____

_____

_____

_____

_____

_____

Results Chart

Time	Control Group	Test Group 1	Test Group 2	Test Group 3	Test Group 4

Written Assignment Week 11

Discussion Questions

1. How does heat travel?
2. What are the three ways heat can travel? Give a brief description of each.
3. What is the purpose of insulation?
4. How does a vacuum flask prevent the three types of heat travel?

Written Assignment Week 11

Student Assignment Sheet Week 12
Engines

Science Fair Project

This week, you will complete steps seven and eight of your Science Fair Project. You will be writing and preparing a presentation of your Science Fair Project.

7. **Create a Board** – This week, you will be creating a visual representation of your science fair project that will serve as the centerpiece of your presentation. You will begin by planning the look of your board, then move onto preparing the information, and finally you will pull it all together.

8. **Give a Presentation** – After you have completed your presentation board, determine if you would like to include part of your experiment in your presentation. Then, prepare a five minute talk about your project, be sure to include the question you tried to answer, your hypothesis, a brief explanation of your experiment and the results plus the conclusion to your project. Be sure to arrive on time for your presentation. Set up your project board and any other additional materials. Give your talk and then ask if there are any questions. Answer the questions and end your time by thanking whoever has come to listen to your presentation.

Vocabulary & Memory Work

☐ Vocabulary: internal combustion engine, power, steam engine

☐ Memory Work—This week, continue to work on memorizing the laws of thermodynamics and the power equation.

⚡ Power (P) = Work (W) / Time (t)

Sketch

▨ There is no sketch this week, to allow for more time to work on the science fair project.

Writing

↝ Reading Assignment: *DK Encyclopedia of Science* pp. 143-144 (Engines)

↝ Additional Research Reading

📖 Combustion: *KSE* pp. 254

Dates to Enter

🕐 1712 – The first steam engine is built by Thomas Newcomen.

🕐 1765 – James Watt improves upon the original Newcomen steam engine.

🕐 1860 – The first internal combustion engine is built by Etienne Lenoir. He uses coal gas and air for fuel.

🕐 1884 – The first steam engine to generate electricity is invented by Charles Parsons.

🕐 1926 – The first rocket propelled by liquid fuel is launched by Robert Goddard.

Science Fair Project Step 7: Create a Board

Key 1 ~ Plan out the board.

Use the template on the following page to sketch out a rough plan for your presentation board.

Key 2 ~ Prepare the information.

Type up the following information for your presentation board:

- ☐ Introduction
- ☐ Hypothesis
- ☐ Research
- ☐ Materials
- ☐ Procedure
- ☐ Results
- ☐ Conclusion

Key 3 ~ Put the board together.

- ☐ Put the decorative elements on your project board.

- ☐ Print out and attach your information paragraphs.

- ☐ Add the title to your science fair project board.

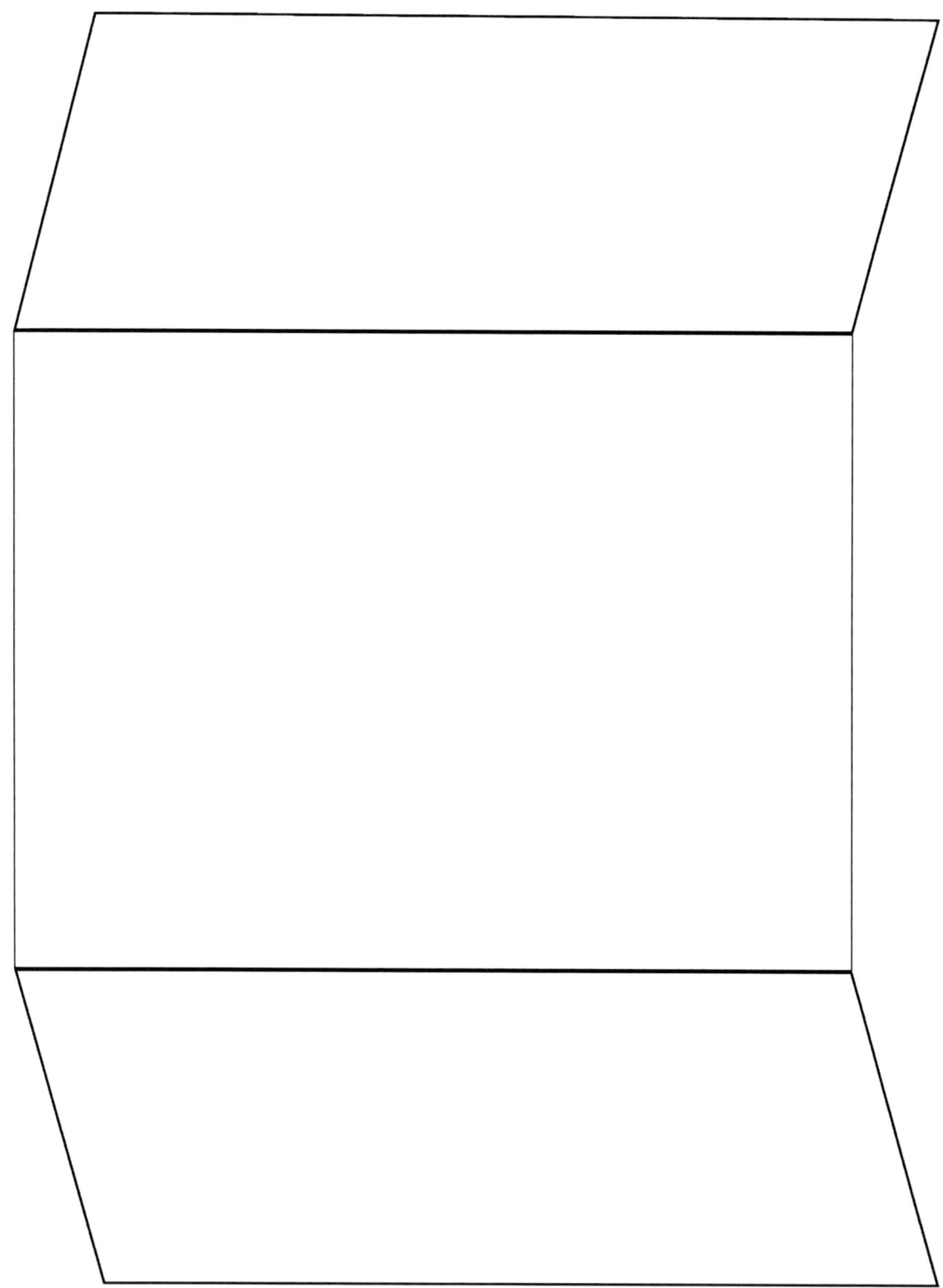

Student Guide Unit 3: Thermodynamics ~ Week 12 Engines

Science Fair Project Step 8: Give a Presentation

Key 1 ~ Prepare the presentation.

Write down the outline for your presentation on a separate sheet of paper.

Key 2 ~ Practice the presentation.

☐ Practice your presentation in front of a mirror several times.

☐ Practice your presentation with your teacher.

Key 3 ~ Share the presentation.

Keep the following tips in mind for your presentation:

- ✓ Arrive on time for your presentation.
- ✓ Set up your project board and any other additional materials.
- ✓ Give your talk and then ask if there are any questions.
- ✓ Answer the questions and end your time by thanking whomever has come to listen to your presentation.

Written Assignment Week 12

Discussion Questions

1. What does every engine do?
2. What is the main difference between an internal combustion engine and a steam engine?
3. How does an internal combustion engine work?
4. How does a steam engine work?
5. How do rockets work?
6. What is jet propulsion?

Written Assignment Week 12

Physics
Unit 4

Sound

Unit 4 Sound
Vocabulary Sheet

Define the following terms as they are assigned on your Student Assignment Sheet.

1. Mechanical wave – _____

   _____

   _____

2. Sound – _____

   _____

   _____

3. Vibration – _____

   _____

   _____

4. Amplitude – _____

   _____

   _____

5. Frequency – _____

   _____

   _____

6. Wavelength – _____

   _____

   _____

7. Antinoise – _____

   _____

   _____

8. Resonate frequency – _____

   _____

   _____

9. Acoustics – _____

_____

_____

10. Echolocation – _____

_____

_____

Student Assignment Sheet Week 13
Sound

Experiment: Does sound travel in a vacuum?

Materials
- ✓ Glass bottle
- ✓ Bell
- ✓ Cork that fits the top of the glass bottle
- ✓ Thread
- ✓ Needle
- ✓ Match

Procedure
1. Read the introduction to the experiment and answer the question for the hypothesis section.
2. Thread the needle and then pass the thread through the bell and up through the cork, so that it will hang inside the bottle and swing freely.
3. Lightly place the cork in top of the bottle and shake gently. Observe if you hear any noise and write your observations on your experiment sheet.
4. Next, remove the cork. Light a match and drop it into the bottle. Quickly replace the cork and make sure it is firmly in place. Set the bottle on the counter and wait for the match to burn out.
5. Once the bottle is cool to the touch, gently shake it and observe if you hear any noise. Write your observations on your experiment sheet.
6. Draw conclusions and complete the experiment sheet.

Vocabulary & Memory Work
- ☐ Vocabulary: mechanical wave, sound, vibration
- ☐ Memory Work—This week, begin working on memorizing the types of mechanical waves.
  - ↳ Transverse Wave – A wave that causes the medium to vibrate perpendicular to the direction in which the wave travels.
  - ↳ Longitudinal Waves – A wave that causes the medium to vibrate parallel to the direction in which the wave travels.
  - ↳ Surface Waves – A wave that travels along the surface separating two medias.

Sketch: Transverse and Longitudinal Waves
- ▦ Label the following – transverse wave, longitudinal wave.
- ▦ Add a double arrow to show the direction of the original movement.

Writing
- ✍ Reading Assignment: *DK Encyclopedia of Science* pp. 178-179 (Sound)
- ✍ Additional Research Readings
  - 📖 Waves: *UIDS* pp. 34-35
  - 📖 Sound as Changes of Pressure: *KSE* pp. 312-313

Dates
- ⏲ 1708 – William Derham successfully establishes the speed of sound.
- ⏲ 1890's – Ernst Mach describes how shock waves form and, along with his son Ludwig, develops a way to take pictures of the shadow of an invisible shock wave.

Sketch Week 13

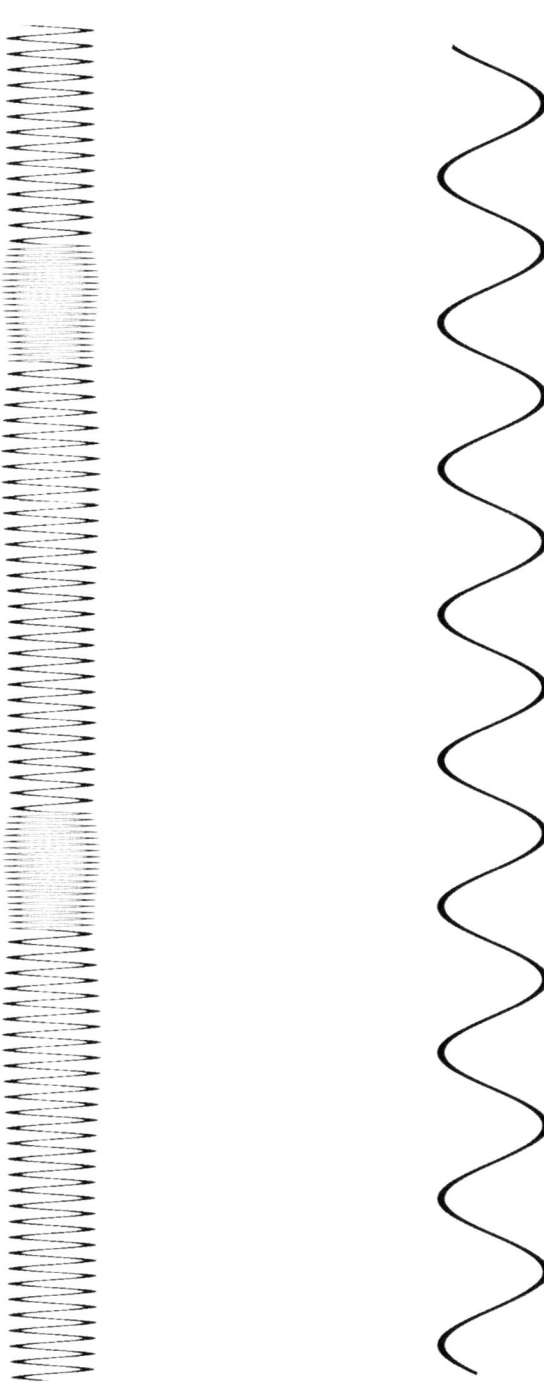

Does sound travel in a vacuum?

Introduction

Sound is a mechanical wave or vibration that can be heard. Both animals and humans have special organs, known as ears, which are able to detect the vibrations caused by sound waves. The human ear can hear sound when an object vibrates at sixteen times per second and as high as twenty thousand times per second. In today's experiment, you are going to test whether the human ear can detect sound when the object is in a vacuum.

Hypothesis

_____

_____

_____

Materials

_____   _____

_____   _____

_____   _____

_____   _____

_____   _____

Procedure

_____

_____

_____

_____

_____

_____

_____

_____

_____

Observations and Results

Bell in the Bottle	Can you hear the bell ringing?
Normal (with a loose cork and the normal amount of air present)	
In a Vacuum (with a tight cork and all the air removed)	
Optional - Underwater (with a tight cork and filled with water)	

_____

_____

_____

_____

_____

_____

_____

Conclusion

_____

_____

_____

_____

Written Assignment Week 13

Discussion Questions

1. What is sound caused by?
2. What are two ways that sound travels differently through water than air?
3. Why does sound travel more quickly through a solid than a gas?
4. What causes a sonic boom?

Written Assignment Week 13

Student Assignment Sheet Week 14
Sound Waves

Experiment: Does distance affect the number of vibrations experience by a medium?

Materials
- ✓ Shallow glass bowl or cup
- ✓ Water
- ✓ Music Player
- ✓ Measuring tape

Procedure
1. Read the introduction to the experiment and answer the question for the hypothesis section.
2. Fill the bowl with water and set it right next to the speaker of the music player. Choose a fast selection of music and play it at a relatively loud volume. Observe the changes in the water in the bowl and sketch what you see.
3. Now, move the player 12 in (30 cm) away from the bowl. Play the same selection of music once more. Observe the changes in the water in the bowl and sketch what you see.
4. Then, move the player 24 in (60 cm) away from the bowl. Play the same selection of music again. Observe the changes in the water in the bowl and sketch what you see.
5. Finally, move the player 36 in (90 cm) away from the bowl. Play the same selection of music for the final time. Observe the changes in the water in the bowl and sketch what you see.
6. Draw conclusions and complete the experiment sheet.

Vocabulary & Memory Work
- ☐ Vocabulary: amplitude, frequency, wavelength
- ☐ Memory Work—This week, continue to work on memorizing the types of mechanical waves along with the speed of waves equation:
  - ♪ Speed (v) = Wavelength (λ) • Frequency (f)

Sketch: Sound Wave
- 🖼 Label the following – crest, trough, wavelength, amplitude.

Writing
- ᕙ Reading Assignment: *DK Encyclopedia of Science* pg. 180 (Measuring Sound), pg. 188 (Sound Recording)
- ᕙ Additional Research Readings
  - 📖 Wave Motion: *KSE* pp. 314-315
  - 📖 Vibrations: *KSE* pp. 316-317
  - 📖 Sound Waves: *UIDS* pp. 40-41

Dates
- 🕐 1877 – The first sound recording is made by Thomas Edison on a phonograph through the mechanical vibrations of a needle.
- 🕐 1887 – Henrich Hertz proves that electricity can be transmitted in electromagnetic waves. He was also the first to produce and detect radio waves.

Sketch Week 14

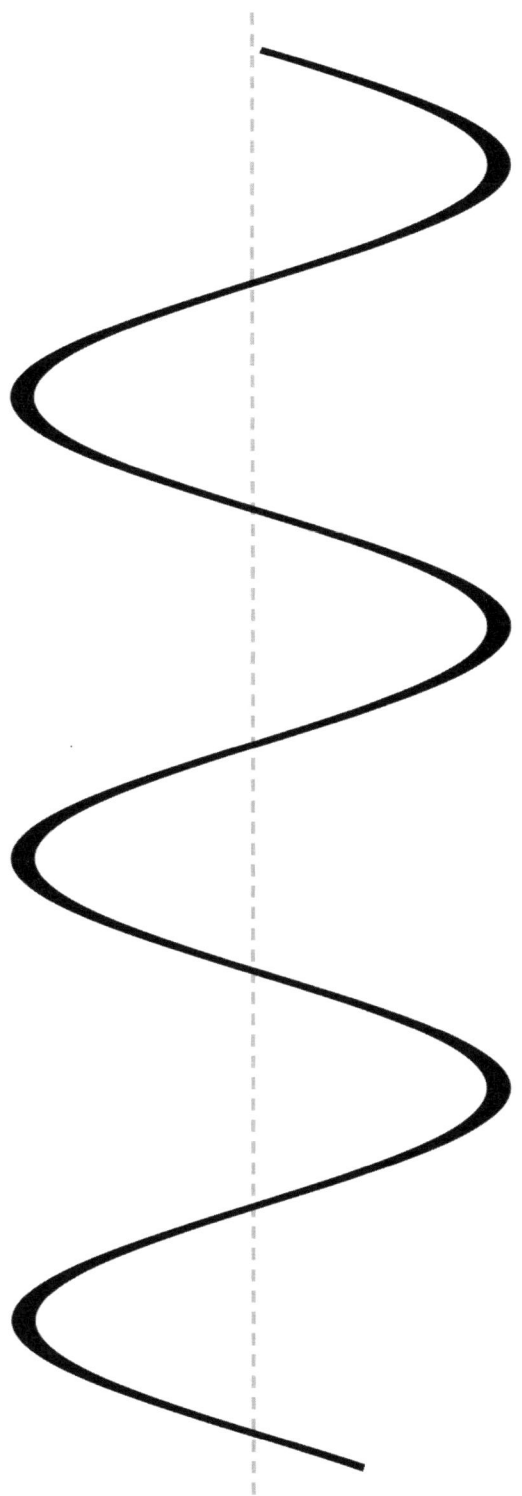

Does distance affect the number of vibrations experience by a medium?

Introduction

As mechanical wave is produced when a source of energy creates a vibration that travels through a medium, such as air or water. A vibration is basically repeated back-and-forth motion, and it can be seen on the surface of a medium such as water. As the sound waves move through the water, a vibration visually disturbs the surface. The greater the strength of the sound wave, the more the water is disturbed. In today's experiment, you are going test if the distance from the energy source makes a different in amount of vibrations that are transferred through the water.

Hypothesis

_____

_____

_____

Materials

_____   _____

_____   _____

_____   _____

Procedure

_____

_____

_____

_____

_____

_____

_____

_____

_____

_____

Observations and Results

_____

_____

Next to the bowl	12 in" from the bowl

24 in" from the bowl	36 in" from the bowl

Conclusion

_____

_____

_____

_____

Written Assignment Week 14

Discussion Questions

Measuring Sound, pg. 180

    1. Does sound travel at different speeds?

    2. What determines the loudness of a sound?

    3. What causes the pitch of a sound to change?

Sound Recording, pg. 188

    1. What are sound recordings?

    2. What is the difference between analog and digital recordings?

Written Assignment Week 14

Student Assignment Sheet Week 15
Hearing Sound

Experiment: Make a Tonoscope

Materials
- ✓ Plastic jar, or small flower pot
- ✓ A piece of latex material large enough to cover the lid of your jar (like the kind used for exercise bands)
- ✓ 1" plastic tubing
- ✓ Rubber band
- ✓ Air-dry clay
- ✓ Salt

Procedure
1. Read the introduction to the experiment.
2. Cut a hole in the lower half of the plastic jar large enough to fit the tubing. Insert the tubing into the hole you just made. Then, pack a bit of clay around the edges so that air won't leak out.
3. Next, stretch the latex material over the opening of the jar and use the rubber band to secure it in place. Sprinkle a bit of salt over the rubber material.
4. Hum, sing, or speak through the tubing and watch what happens to the salt. Change the pitch of your voice and observe any changes. Write your observations on your experiment sheet.
5. Draw conclusions and complete the experiment sheet.

Vocabulary & Memory Work
- ☐ Vocabulary: antinoise, resonate frequency
- ☐ Memory Work—This week, continue working on memorizing the types of mechanical waves.

Sketch: Sound Production
- ▨ Draw a line representing the range of sounds the human, monkey, frog, and grasshopper can make. Label the lines with the following – 85 – 1,100 Hz (Human), 400 – 6,000 (Monkey), 50 – 8,000 Hz (Frog), and 7,000 – 100,000 Hz (Grasshopper).

Writing
- ᔑ Reading Assignment: *DK Encyclopedia of Science* pg. 181 (Loudness), pp. 182-183 (Making and Hearing Sound)
- ᔑ Additional Research Readings
  - ☐ Electronic Sound: *DK EOS* pp. 189
  - ☐ Resonance: *KSE* pp. 318-319

Dates
- ☉ 6th century BC – Pythagoras directly links the amplitude of the vibration of a plucked string to the perceived loudness of the instrument.

Sketch Week 15

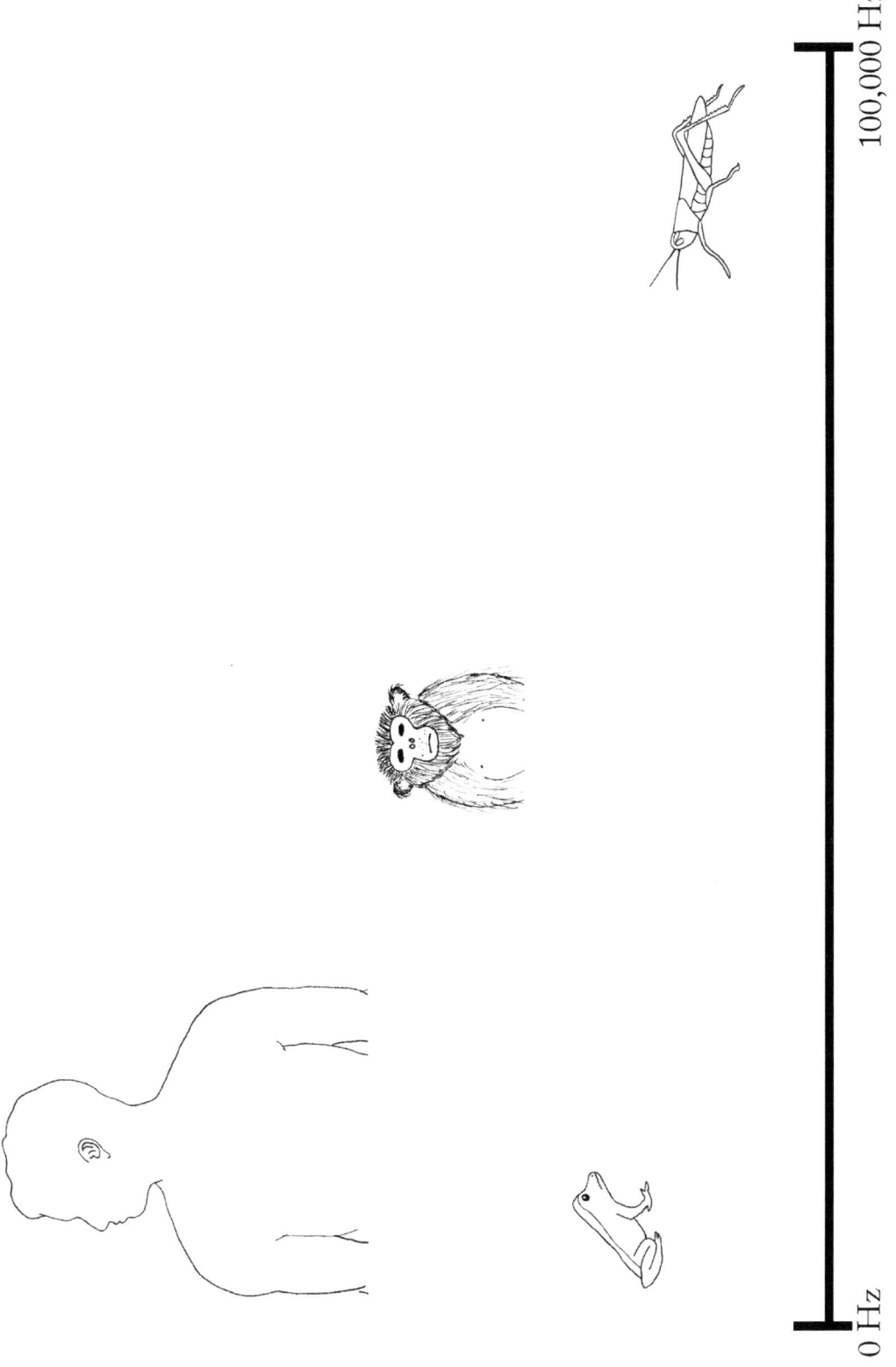

100,000 Hz

0 Hz

Make a Tonoscope

Introduction

A tonoscope is an acoustically device that allows you to see patterns in salt created by the sound of your voice. It was originally designed by Hans Jeny to explore wave behavior and patterns. In today's experiment, you are going to make your own tonoscope.

Materials

_____   _____

_____   _____

_____   _____

Procedure

_____

_____

_____

_____

_____

_____

_____

_____

_____

_____

_____

Observations and Results

_____

_____

_____

_____

My Tonoscope

_____

_____

_____

Conclusion

_____

_____

_____

_____

_____

Written Assignment Week 15

Discussion Questions

Loudness , pg. 181

    1. What does the loudness of a sound depend upon?

    2. What does the decibel scale measure?

Making and Hearing Sound, pp. 182-183

    1. How do we hear sound?

    2. What is resonance?

    3. How does a voice activated device work?

Written Assignment Week 15

Student Assignment Sheet Week 16
Acoustics

Experiment: Does the size of the room I am in affect the way I hear?

Materials
- ✓ Partner
- ✓ Blindfold

Procedure

1. Read the introduction to the experiment and answer the question for the hypothesis section.
2. Find the largest room in your house, like the basement or garage. Stand in the center of the room and put on the blindfold. Have your partner stand in front, in back, or to the side of you. He or she should be about 3 feet (1 meter) from you with their back to you. Have your partner whisper your name and then try to determine his or her location. Repeat this process until your partner has stood in front, in back, and to both sides of you. Write down on your experiment sheet which locations you got correct.
3. Next find an average-sized room in your house, like a bedroom. Repeat the process from step 2.
4. Finally find the smallest room in your house, like a bathroom. Repeat the process from step 2.
5. Draw conclusions and complete the experiment sheet.

Vocabulary & Memory Work

- ☐ Vocabulary: acoustics, echolocation
- ☐ Memory Work—This week, continue working on memorizing the types of mechanical waves.

Sketch: Dolphin Echolocation

- Label the following – dolphin makes a clicking sound, sending energy into the surrounding water; soft surfaces absorb part of the sound energy and reflect the rest; hard surfaces reflect the sound energy; energy travels through the water

Writing

- ✐ Reading Assignment: *DK Encyclopedia of Science* pp. 184-185 (Reflection and Absorption)
- ✐ Additional Research Readings
  - 📖 Musical Sounds: *DK EOS* pp. 186-187
  - 📖 Vibrations of Strings: *KSE* pg. 320
  - 📖 Vibrations in Tubes: *KSE* pg. 321
  - 📖 Perception of Sound: *UIDS* pp. 42-43

Dates

- 🕐 1906 – American naval architect Lewis Nixon invents the first sonar-like listening device, which he used to detect icebergs.

Sketch Week 16

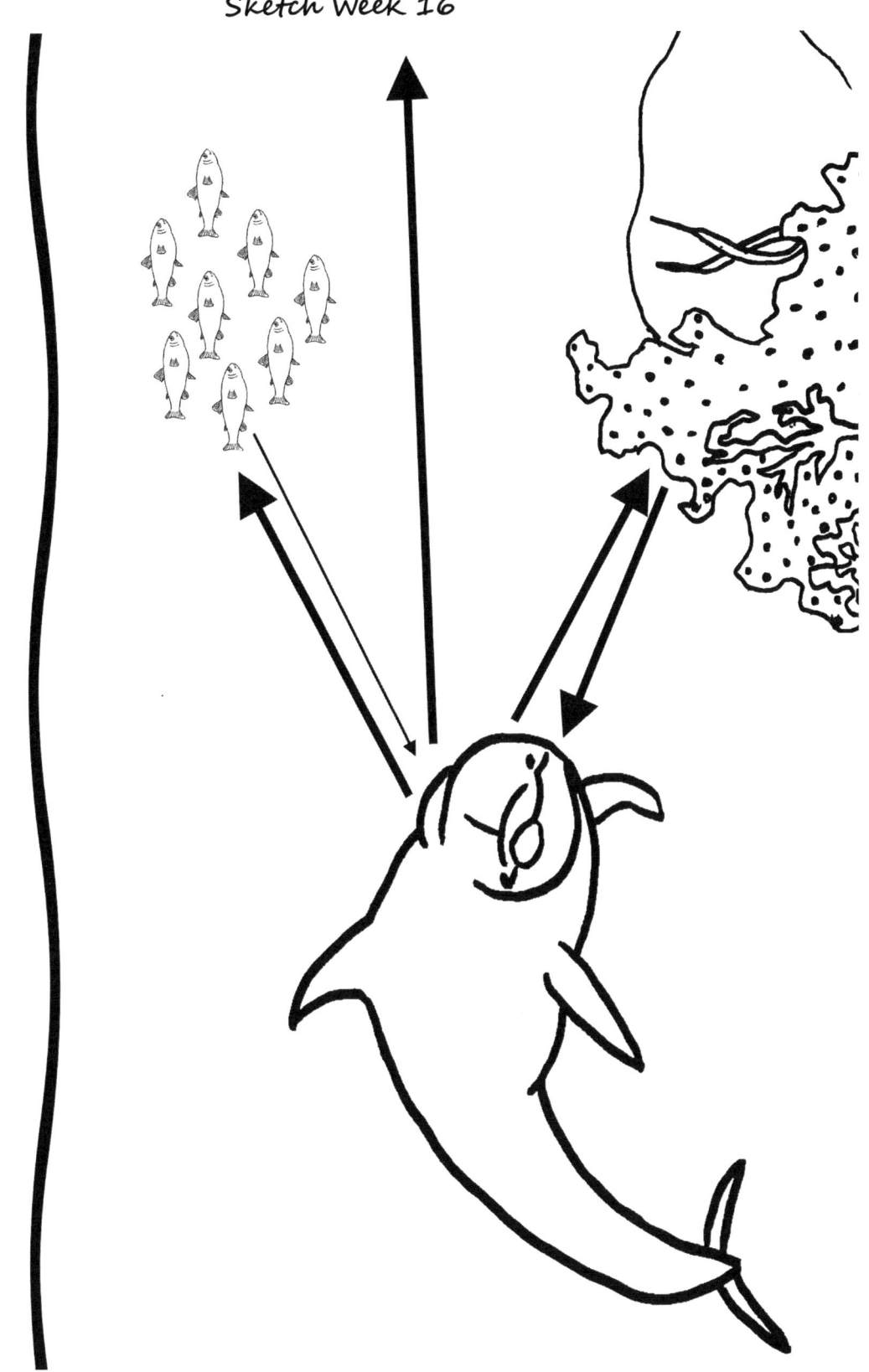

Does the size of the room I am in affect the way I hear?

Introduction

Acoustics is the study of sound and the way it travels in a given space. When designing a performance hall, architects will use certain materials in specific places to create the desired sound effects for the space. In today's experiment, you are going to test the acoustics of various rooms in and outside your home.

Hypothesis

_____

_____

_____

Materials

_____   _____

_____   _____

_____   _____

Procedure

_____

_____

_____

_____

_____

_____

_____

Observations and Results

Room	In Front	Behind	On Left	On Right
Large Room				

Middle-sized Room				
Smaller Room				

_____

_____

_____

Conclusion

_____

_____

_____

_____

_____

Written Assignment Week 16

Discussion Questions

1. What happens when sound waves hit a hard surface? A soft surface?
2. What are echoes?
3. What does a parabolic dish do?
4. How do dolphins and other animals use echolocation?
5. How does ultrasound and sonar work?

Written Assignment Week 16

Physics
Unit 5

Light

Unit 5 Light
Vocabulary Sheet

Define the following terms as they are assigned on your Student Assignment Sheet.

1. Electromagnetic waves – _____

_____

_____

2. Electromagnetic spectrum – _____

_____

_____

3. Photons – _____

_____

_____

4. Reflection – _____

_____

_____

5. Refraction – _____

_____

_____

6. Convex – _____

_____

_____

7. Concave – _____

_____

_____

8. Interference – _____

_____

_____

9. Diffraction – _____

_____

_____

10. Converging lens – _____

_____

_____

11. Diverging lens – _____

_____

_____

12. Lens – _____

_____

_____

Student Assignment Sheet Week 17
Light

Experiment: Does sunscreen really block UV rays?

Materials

- ✓ 9 Ultraviolet light detecting beads
- ✓ 3 Shallow dishes (not clear plastic or glass)
- ✓ Plastic Wrap
- ✓ Two different levels of SPF sunscreen (i.e., SPF 15 and SPF 45)
- ✓ Rubber bands

Procedure

1. Read the introduction to the experiment and answer the question for the hypothesis section.
2. Place three of the ultraviolet detecting beads in each of the bowls and label them a "A," "B," and "C." Then, cut the plastic wrap to cover bowl each one and secure it in place with rubber band or tape.
3. Next, slather a fair amount of the lower SPF sunscreens over the plastic wrap paper on bowl "B," and a fair amount of the higher SPF sunscreen over the plastic wrap on bowl "C." Place on three bowls on a tray, take them outside, and set them on a table in the full sun.
4. Check the beads after 30 seconds, 2 minutes, 5 minutes, 10 minutes, and 30 minutes. Each time, rate the degree of color change on a scale of 1 to 10 (e.g. If you beads have changed color by about 20%, rate them a 2). Record your observations each time on your experiment sheet.
5. Draw conclusions and complete the experiment sheet.

Vocabulary & Memory Work

- ☐ Vocabulary: electromagnetic waves, electromagnetic spectrum, photons
- ☐ Memory Work—This week, begin working on memorizing the waves of the electromagnetic spectrum and the speed of waves equation (if you did not memorize it in the last unit).

Sketch: Light

- 🖾 Label the following – source of light; at times, light behaves as if it is composed of a stream of particles; at times, light behaves as if it is a wave motion.

Writing

- ᙰ Reading Assignment: *DK Encyclopedia of Science* pp. 190-191 (Light), pg. 192 (Electromagnetic Spectrum)
- ᙰ Additional Research Readings
  - 📖 Sources of Light: *DK EOS* pg. 193
  - 📖 Light and Matter: *DK EOS* pg. 200
  - 📖 Light: *KSE* pp. 260-261, *UIDS* pg. 46

Dates

- 🕐 1672 – Isaac Newton suggests that light is composed of tiny particles resembling balls.
- 🕐 1678 – Christian Huygens suggests that light is a wave motion, similar to sound or water waves.
- 🕐 1900 – Max Planck suggests that light is a combination of a particle and a wave, forming the basis of quantum theory.
- 🕐 1905 – Einstein expands upon Planck's work and suggests that light is composed of tiny particles, called photons, which have energy and behave like waves.

Sketch Week 17

Does sunscreen really block UV rays?

Introduction

Ultraviolet light from the Sun is beneficial. These electromagnetic waves are necessary for our skin to produce Vitamin D, and the waves help plants to grow. However, prolonged exposure to ultraviolet light rays can cause sunburn, wrinkles, and eventually skin cancer. In today's experiment, you are going to test to see whether or not sunscreen helps to block Ultraviolet light rays.

Hypothesis

_____

_____

_____

Materials

_____   _____

_____   _____

_____   _____

_____   _____

_____   _____

Procedure

_____

_____

_____

_____

_____

_____

_____

_____

_____

_____

Observations and Results

Jar	30 seconds	2 minutes	5 minutes	10 minutes	30 minutes
Jar with no cover					
Jar covered with plastic wrap					
Jar covered with plastic wrap and sunscreen					

_____

_____

_____

_____

_____

_____

_____

Conclusion

_____

_____

_____

_____

Written Assignment Week 17

Discussion Questions

Light, pp. 190-191

   1. What is light?

   2. How does light differ from sound?

   3. What is the photoelectric effect?

   4. What does quantum theory say about light?

Electromagnetic Spectrum, pg. 192

   1. How are electromagnetic waves similar? How do they differ?

   2. Which electromagnetic waves have shorter wavelengths and carry more energy that visible light? Which ones have longer wavelengths and carry less energy than visible light?

Written Assignment Week 17

Student Assignment Sheet Week 18
Reflection and Refraction

Experiment: Which liquid splits a pencil the most?

Materials
- ✓ 4 Pencils
- ✓ 4 Clear glasses
- ✓ Water
- ✓ Oil
- ✓ Alcohol
- ✓ Corn syrup

Procedure
1. Read the introduction to the experiment and answer the question for the hypothesis section.
2. Label the cups #1 through #4 and place a pencil in each cup. Observe how the pencil looks in the cup without any liquid.
3. Now add ½ cup (about 120 mL) of water to cup #1, add ½ cup (about 120 mL) of oil to cup #2, ½ cup (about 120 mL) of alcohol to cup #3, and ½ cup (about 120 mL) of corn syrup to cup #4.
4. Observe what happens to the pencils in the glasses and write your observations down on your experiment sheet.
5. Draw conclusions and complete the experiment sheet.

Vocabulary & Memory Work
- ☐ Vocabulary: reflection, refraction
- ☐ Memory Work—This week, continue to work on memorizing the waves of the electromagnetic spectrum.

Sketch: Reflection
- 🖼 Label the following – Reflection, mirror, angle of incidence, angle of reflection, the law of reflection states that the angle of reflection is equal to the angle of incidence
- 🖼 Draw arrows to show what happens to the light rays when they are reflected and when they are refracted.

Writing
- ⌇ Reading Assignment: *DK Encyclopedia of Science* pp. 194-195 (Reflection), pg. 196 (Refraction)
- ⌇ Additional Research Readings
  - 📖 Reflection: *KSE* pp. 262-263, *UIDS* pp. 47-49
  - 📖 Refraction: *KSE* pp. 264-265, *UIDS* pp. 50-53

Dates
- 🕐 1902 – Hendrick Lorentz is awarded the Nobel Peace Prize for his work with electromagnetic waves and the propagation of light.

Sketch Week 18

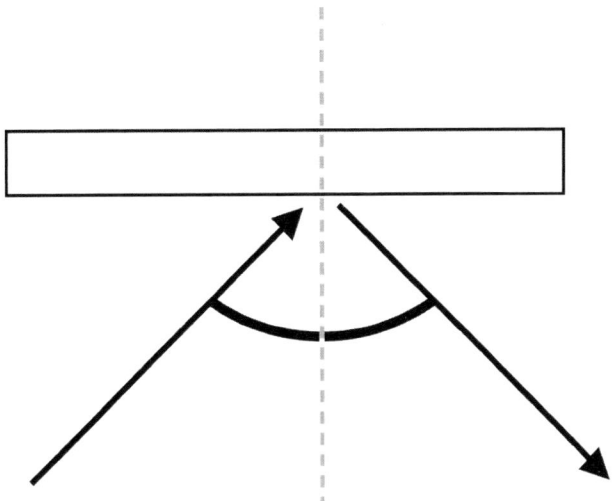

Which liquid splits a pencil the most?

Introduction

Light typically travels in a straight line. However, as it passes through different media, it bends slightly. This phenomenon is known as refraction and it occurs because light travels at different speeds through different media. In today's experiment, you are going to test how different liquids refract the reflected light from a pencil is refracted in different liquids.

Hypothesis

_____

_____

Materials

_____   _____

_____   _____

_____   _____

Procedure

_____

_____

_____

_____

_____

_____

_____

_____

_____

Observations and Results

_____

_____

_____

Cup with Water	Cup with Oil

Cup with Alcohol	Cup with Corn Syrup

Conclusion

_____

_____

_____

_____

_____

_____

_____

_____

Written Assignment Week 18

Discussion Questions

Reflection, pp. 194-195

1. How do we see objects that don't make their own light?
2. What happens when a surface does not reflect light?
3. What causes a mirror image?
4. What kind of images do convex mirrors produce? Concave mirrors?
5. What is the law of reflection?

Refraction, pg. 196

1. What happens when light travels from one transparent material to another?
2. What is the refractive index of a material?

Written Assignment Week 18

Student Assignment Sheet Week 19
Vision and Color

Experiment: Can I trick my brain and change the color?

Materials
- ✓ Thin cardboard
- ✓ Red, blue, and yellow paint
- ✓ 6 Rubber bands
- ✓ Hole punch

Procedure
1. Read the introduction to the experiment and answer the questions in the hypothesis section.
2. Cut out 6 circles of the same size from the cardboard. Paint two of them red, two of them yellow, and two of them blue. Set them aside to dry.
3. Once they are dry, glue one of the red disks to one of the yellow disks, one of the red disks to one of the blue disks, and one of the blue disks to one of the yellow disks. Then, punch holes on opposite sides of the three disks and tie a rubber band into each hole.
4. Take the red/blue disk, twist the rubber bands, and then stretch them out so that the disk spins quickly. Observe what color you see and write it on your experiment sheet.
5. Repeat step 4 with the red/yellow and blue/yellow disks.
6. Draw conclusions and complete the experiment sheet.

Vocabulary & Memory Work
- ☐ Vocabulary: convex, concave, diffraction, interference
- ☐ Memory Work—This week, continue to work on memorizing the waves of the electromagnetic spectrum.

Sketch: Spectrum of Visible Light
- 🔲 Label the following – Visible light; Prism splits visible light into its spectral components; Red light, Wavelength 610 to 750 nm; Orange light, Wavelength 590 to 610 nm; Yellow light, Wavelength 570 to 590 nm; Green light, Wavelength 500 to 570 nm; Blue light, Wavelength 450 to 500 nm; Violet light, Wavelength 400 to 450 nm.

Writing
- ᕳ Reading Assignment: *DK Encyclopedia of Science* pg. 202 (Color), pp. 204-205 (Vision)
- ᕳ Additional Research Readings
  - 📖 Color Subtraction: *DK EOS* pg. 203
  - 📖 Color: *KSE* pp. 272-273

Dates
- 🕐 There are no dates to be entered this week.

Sketch Week 19

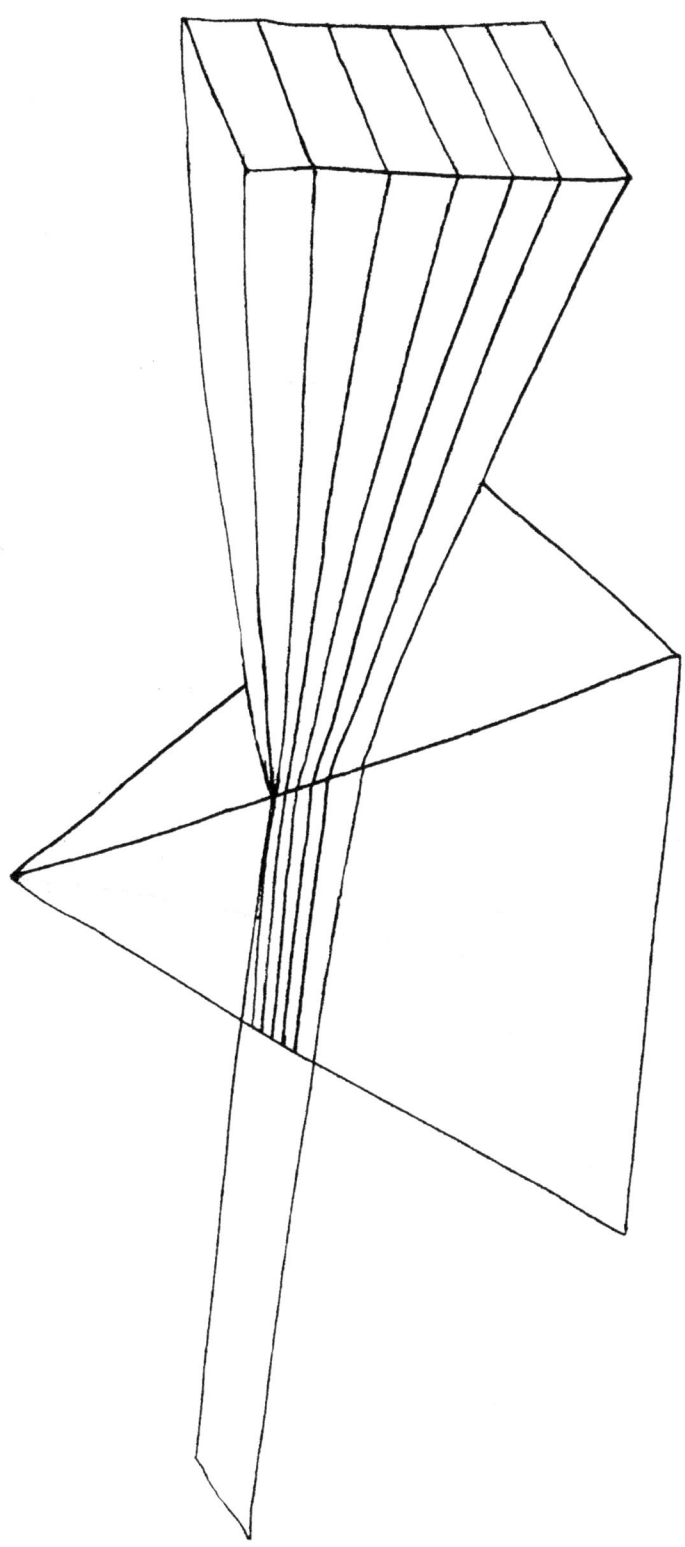

Can I trick my brain and change the color?

Introduction

Visible light is made up of a spectrum of wavelengths. When we see color, we are actually see the wavelength(s) of light that are reflected by the object. Light receptors in our eyes receive the reflected wavelengths and then our brain translated into the sensation of color. In today's experiment, you are going to try to trick your brain into seeing different colors than are actually there.

Hypothesis

_____

_____

_____

Materials

_____   _____

_____   _____

_____   _____

Procedure

_____

_____

_____

_____

_____

_____

_____

_____

_____

_____

_____

_____

Observations and Results

	Red and Blue Disks	Red and Yellow Disks	Blue and Yellow Disks
Color(s) I saw			

_____

_____

_____

_____

_____

_____

_____

Conclusion

_____

_____

_____

_____

_____

Written Assignment Week 19

Discussion Questions

Color, pg. 202
1. What makes a color?
2. What happens when all the wavelengths of visible light are mixed?
3. What happens in a prism?

Vision, pp. 204-205
1. How do the eyes and brain work together to allow us to see?
2. How can a concave lens help to correct vision? A convex one?
3. Why do we see colors differently during the day and night?

_____

_____

_____

_____

_____

_____

_____

_____

_____

_____

_____

_____

_____

_____

_____

_____

_____

_____

_____

_____

Written Assignment Week 19

Student Assignment Sheet Week 20
Optics

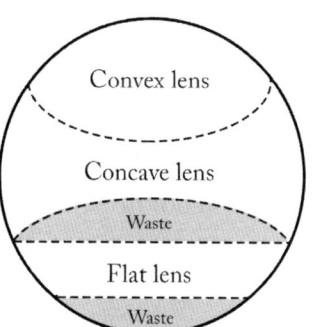

Experiment: How does shape affect the way a lens projects light?
Materials
- ✓ 1 Package Jell-O™ (orange, lemon, or lime)
- ✓ Round, flat-bottomed bowl or jar – at least 4" (10 cm) in diameter
- ✓ 1 Cup water, Dull knife, Plate, Flashlight

Procedure

Note—The night before you do this experiment, mix the package of Jell-O with 1 cup (240 mL) of very hot water. Pour a 1½ inch (about 4 cm) layer of the Jell-O into the round, flat-bottomed jar sprayed with a bit of oil. Set it in the fridge over night to set.

1. Read the introduction to the experiment and answer the question for the hypothesis section.
2. Take the jar out and set it in warm water for 30 seconds to a minute. Gently remove the Jell-O disk and set it on a plate. Cut it into three lenses (concave, flat, and convex) using the diagram above.
3. Move to an interior room with no windows and place the plate about a foot away from the wall.
4. Set the flat lens on the edge of the plate, place the small flashlight behind it, turn the flashlight on, and then turn out the lights. Observe the light pattern that is displayed and measure from one end of the display to the other. Turn the light back on and record this information on your experiment sheet.
5. Next, set the concave lens on the edge of the plate so that the curved part faces the flashlight, turn the flashlight on, and then turn out the lights. Observe the light pattern that is displayed and measure from one end of the display to the other. Turn the light back on and record this information on your experiment sheet.
6. Repeat step 6 with the convex lens. Draw conclusions and complete the experiment sheet.

Vocabulary & Memory Work
- ☐ Vocabulary: converging lens, diverging lens, lens

Sketch: Convex vs. Concave
- ▨ Label the following – concave lens, convex lens, light.
- ▨ Draw the rays of light as they enter and exit the concave and convex lenses.

Writing
- ✐ Reading Assignment: *DK Encyclopedia of Science* pg. 197 (Lenses), pg. 198 (Optical Instruments)
- ✐ Additional Research Readings
  - 📖 Lenses and Curved Mirrors: *KSE* pp. 266-267
  - 📖 Optical Instruments: *UIDS* pp. 54-55

Dates
- 🕐 Late 1600's – Antoni Van Leeuwenhoek designs the first microscope.
- 🕐 1789 – Will Herschel designs a telescope with a four foot diameter.
- 🕐 1992 – The Keck telescope is built, which is thirty-three feet in diameter.

Sketch Week 20

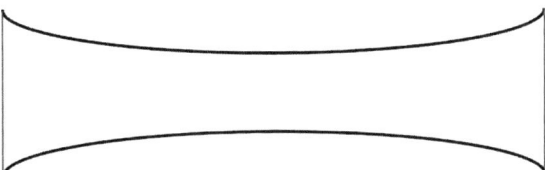

How does shape affect the way a lens projects light?

Introduction

A lens is a piece of transparent material that causes light to bend in a particular way. Lenses are used in glasses, photography, telescopes, and more. They are used to produce images, to magnify objects, to focus light, and to reduce images in a scene. In today's experiment, you are going to examine how the shape of a lens affects how it projects light.

Hypothesis

_____

_____

_____

Materials

_____   _____

_____   _____

_____   _____

Procedure

_____

_____

_____

_____

_____

_____

_____

_____

_____

_____

_____

Observations and Results

	Flat Lens	Convex Lens	Concave Lens
Width of Light Pattern			

_____

_____

_____

Conclusion

_____

_____

_____

_____

_____

Written Assignment Week 20

Discussion Questions

Lenses, pg. 197

1. What fact about light do lenses take advantage of?
2. How does the shape of the lens matter?

Optical Instruments, pg. 198

1. What do optical instruments do?
2. How does a compound microscope work?
3. What is the difference between reflecting and refracting telescopes?

_____

_____

_____

_____

_____

_____

_____

_____

_____

_____

_____

_____

_____

_____

_____

_____

_____

_____

_____

_____

_____

_____

Written Assignment Week 20

Physics
Unit 6

Electricity and Magnetism

Unit 6 Electricity and Magnetism
Vocabulary Sheet

Define the following terms as they are assigned on your Student Assignment Sheet.

1. Electric charge – _____

_____

_____

2. Induction – _____

_____

_____

3. Conductor – _____

_____

_____

4. Insulator – _____

_____

_____

5. Semiconductor – _____

_____

_____

6. Anode – _____

_____

_____

7. Cathode – _____

_____

_____

8. Diode – _____

_____

_____

9. Parallel circuit – _____

_____

_____

10. Resistance – _____

_____

_____

11. Series circuit – _____

_____

_____

12. Electrolyte – _____

_____

_____

13. Magnetic field – _____

_____

_____

14. Magnetic pole – _____

_____

_____

15. Electromagnet – _____

_____

_____

16. Electromagnetic force – _____

_____

_____

17. Solenoid – _____

_____

_____

18. Turbine – _____

_____

Student Assignment Sheet Week 21
Electricity

Experiment: Can I transfer an electrical charge?

Materials
- ✓ Styrofoam tray or plate
- ✓ Aluminum pan or pie-plate
- ✓ Wool
- ✓ Plastic tongs
- ✓ Pin

Procedure
1. Read the introduction to the experiment and answer the question for the hypothesis section.
2. Rub the Styrofoam tray (or plate) vigorously with a piece of wool for about two minutes. Set the tray face down on a smooth surface.
3. Use the plastic tongs to set the aluminum pan (or pie-plate) on top of the Styrofoam tray (or plate). Then, gently slide it back and forth several times.
4. Now, use the tongs to pick up the pin. With your hands on the plastic tongs, move the pin close to the pan and observe what happens.
5. Draw conclusions and complete the experiment sheet.

Vocabulary & Memory Work

- ☐ Vocabulary: electrical charge, induction
- ☐ Memory Work—This week, work on memorizing the law of electrostatics and the electrical charge equation.
    - ⚡ Law of Electrostatics – Like charges repel each other and opposite charges attract each other.
    - ⚡ Electrical charge (C) = Current (I) • time (t)

Sketch: Anatomy of a Lightning Strike

- 🖾 Label the following – negative charges collect at the bottom of a storm clouds, positive charges collect on surface of the ground, electrical discharge path

Writing

- ᕫ Reading Assignment: *DK Encyclopedia of Science* pg. 145 Electricity and Magnetism, pp. 146-147 Static Electricity
- ᕫ Additional Research Readings
    - 📖 Electricity: *KSE* pp. 338-339
    - 📖 Static Electricity: *UIDS* pp. 56-57

Dates

- ⏱ 585 BC – Greek philosopher, Thales, inadvertently discovers static electricity when he rubs a piece of amber with fur and observes how the amber now attracts small objects, like feathers.
- ⏱ 1544-1603 – William Gilbert lives. He is known as the father of electricity and magnetism.
- ⏱ 1753 – Benjamin Franklin announces his lightning conductor invention, which he created as a result of his famous kite-flying experiments.
- ⏱ 1784-1789 – Charles Coulomb, a French physicist, writes and proves the law of electrostatics.

Sketch Week 21

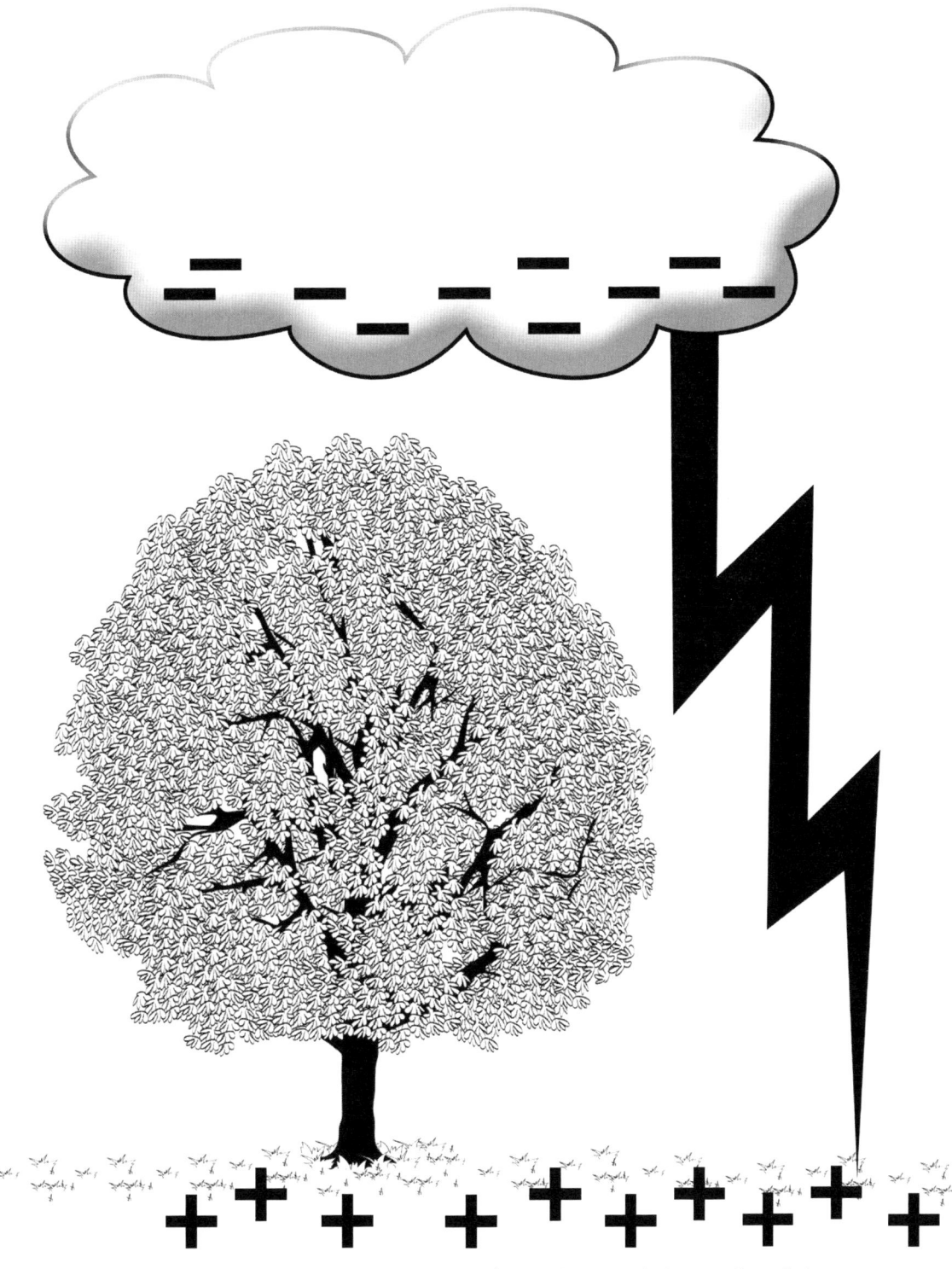

Can I transfer an electrical charge?

Introduction

An electrical charge is produced when a subatomic particle within an object has a shortage or excess of electrons. A shortage of electrons produces a positive electrical charge and an excess of electrons produces a negative electrical charge. The two types of electrical charges are attracted to one another and electricity takes advantage of this principle. In today's experiment, you are going to test whether you can transfer an electrical charge from one object to another.

Hypothesis

_____

_____

_____

Materials

_____   _____

_____   _____

_____   _____

_____   _____

_____   _____

Procedure

_____

_____

_____

_____

_____

_____

_____

_____

_____

Observations and Results

_____

_____

_____

_____

_____

_____

_____

_____

Conclusion

_____

_____

_____

_____

Written Assignment Week 21

Discussion Questions

Electricity and Magnetism, pg. 145

    1. What is electronics?

    2. What is lodestone and what was it used for?

Static Electricity, pp. 146-147

    1. What is static electricity?

    2. What happens when you charge an object through friction?

    3. What is attraction? Repulsion?

    4. What is a capacitor?

Written Assignment Week 21

Student Assignment Sheet Week 22
Conductors and Insulators

Experiment: Can organic materials conduct electricity?

Materials

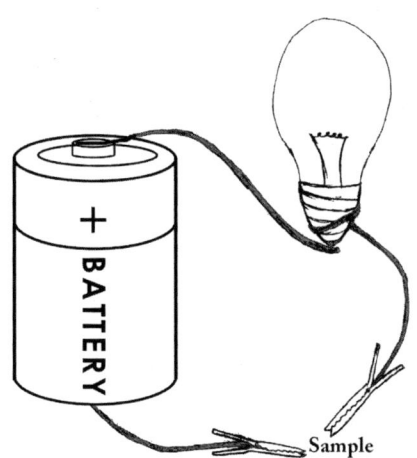

- ✓ Light bulb
- ✓ Copper wire
- ✓ D battery
- ✓ Electrical tape
- ✓ 2 Alligator clips
- ✓ Organic material, such as a pickle, lemon slice, cheese, bread, or a leaf

Procedure

1. Read the introduction to the experiment and answer the question for the hypothesis section.
2. Begin by cutting the wire into three lengths. Then, take one of the wires and attach an alligator clip to one end. Take the next wire and attach one end to the other alligator clip. Then, wrap the other end once around the base of a light bulb.
3. Now, take the two wires with bare ends and attach those to the two terminals of the battery using the electrical tape. (*See the diagram above to visually check your electrical circuit.*)
4. The electrical circuit is now ready for testing. Simply clip the alligator clips on either side of your sample, hold the top of the bulb, and touch the base of the light bulb to the end of the wire coming from the battery to complete the circuit. If the light bulb lights up, the material conducts electricity; if it does not, the sample does not conduct electricity.
5. Draw conclusions and complete the experiment sheet.

Vocabulary & Memory Work

- ☐ Vocabulary: conductor, insulator, semiconductor
- ☐ Memory Work—This week, begin working on memorizing the two types of electrical current and the potential difference equation. (*See Appendix pg. 262 for a complete listing.*)

Sketch: Electron Flow Diagram

- ▨ Label the following – positive terminal, negative terminal, electron flow, current flow
- ▨ Draw the arrows to show the direction of the electron flow and current.

Writing

- ᴧ Reading Assignment: *DK Encyclopedia of Science* pp. 148-149 (Current Electricity)
- ᴧ Additional Research Readings
    - 📖 Conductors: *KSE* pp. 360-361
    - 📖 Insulators: *KSE* pp. 362-363
    - 📖 Semiconductors: *UIDS* pg. 65
    - 📖 Electrical Current: *UIDS* pp. 60-61

Dates

- ⏲ 1987 – The Nobel Prize is awarded to Muller and Bednorz for their work with finding superconductors that function above absolute zero.

Sketch Week 22

Can organic materials conduct electricity?

Introduction

A conductor is a material through which electrical charge can easily flow, while an insulator is a material through which electrical charge cannot easily flow. When we want to move electricity from one place to another, we use copper wire, which easily conducts electrical charge. But what if we could use other renewable organic materials, like pickles, to conduct electricity? In today's lab, you are going to test whether different organic compounds can conduct electricity.

Hypothesis

_____

_____

Materials

_____   _____

_____   _____

_____   _____

Procedure

_____

_____

_____

_____

_____

_____

_____

_____

_____

_____

_____

_____

Observations and Results

Sample Name				
Did it light up the bulb?				

_____

_____

_____

_____

_____

_____

_____

Conclusion

_____

_____

_____

_____

_____

_____

_____

_____

Written Assignment Week 22

Discussion Questions

1. What is electric current and how is it measured?
2. What is the difference between a conductor and an insulator?
3. What does it mean to "dope" a semiconductor and why would you do that?
4. How does electroplating work?

Written Assignment Week 22

Student Assignment Sheet Week 23
Batteries

Experiment: Do dead batteries bounce?

    Materials
- ✓ 2 AA disposable batteries (one fully charged, one completely dead)
- ✓ Ruler

    Procedure
1. Read the introduction to the experiment and answer the question for the hypothesis section.
2. Drop the fully charged battery from a height of 3 inches (7.5 cm). Observe and measure the height of any bounce that occurs. Record the results on your experiment sheet.
3. Drop the completely dead battery from a height of 3 inches (7.5 cm). Observe and measure the height of any bounce that occurs. Record the results on your experiment sheet.
4. Draw conclusions and complete the experiment sheet.

Vocabulary & Memory Work
- ☐ Vocabulary: anode, cathode, diode
- ☐ Memory Work—This week, continue to work on memorizing the law of electrostatics and the two types of electrical current.

Sketch: Anatomy of a dry cell battery
- ▨ Label the following – positive terminal, negative terminal, Zinc casing (negative electrode), Carbon rod (positive electrode), powdered carbon and magnesium oxide, ammonium chloride paste (electrolyte)

Writing
- ᘓ Reading Assignment: *DK Encyclopedia of Science* pp. 150-151 (Cells and batteries)
- ᘓ Additional Research Readings
  - ▢ Electrochemistry: *KSE* pg. 356
  - ▢ Power Cells: *KSE* pg. 357
  - ▢ Cells and batteries: *UIDS* pp. 68-69

Dates
- ☽ 1800 – Italian scientist, Volta, invents the first battery able to hold an electrical charge.

Sketch Week 23

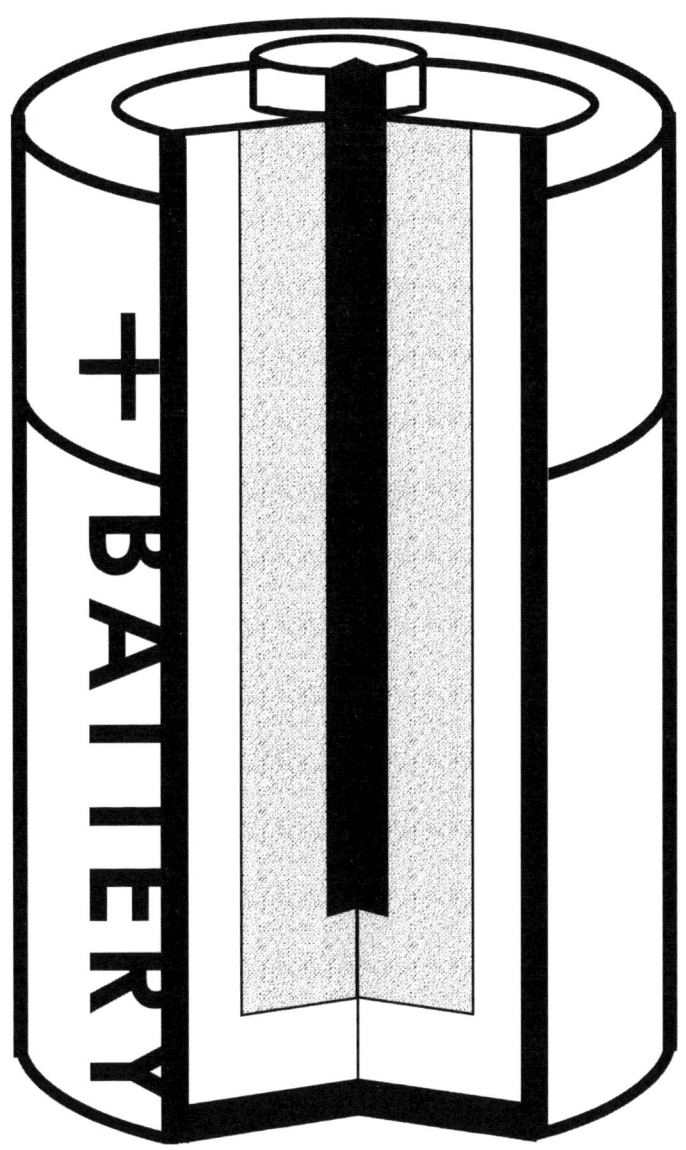

Do dead batteries bounce?

Introduction

Batteries hold electrical charge that is used to power many of our modern-day portable electronics. For years, there has been an old-wives' tale that you can test if a battery is dead by dropping it. If it bounces, it is dead. If not, it still holds a charge. In today's experiment, you are going to test to see if this myth is true.

Hypothesis

_____

_____

_____

Materials

_____   _____

_____   _____

_____   _____

Procedure

_____

_____

_____

_____

_____

_____

_____

_____

_____

_____

_____

_____

_____

Observations and Results

	Fully Charged Battery	Completely Dead Battery
Height of Bounce		

_____

_____

_____

_____

_____

_____

_____

Conclusion

_____

_____

_____

_____

_____

Written Assignment Week 23

Discussion Questions

1. What do all batteries have in common?
2. What is the difference between a cell and a battery?
3. What are the three basic components of a cell?
4. What is the benefit of a larger cell battery?
5. How does a solar cell work?

Written Assignment Week 23

Student Assignment Sheet Week 24
Circuits

Experiment: Online Circuits Lab

Materials

✓ Computer with Internet connection

Procedure

1. Before you begin the lab, draw a circuit that you think will work under the hypothesis section on your experiment sheet.
2. Begin the online lab by clicking on the website below to begin your lab:
   ☉ http://phet.colorado.edu/en/simulation/circuit-construction-kit-dc-virtual-lab
   Note—If you cannot get the link to work, head to PhET's main website at http://phet. colorado.edu/. Enter "circuits" into the search box and then click on the link for the "Circuit Construction Kit (DC Only)."
3. Click on the "Run Now" button and then use the available materials in the box to test the circuit you designed. Record on your experiment sheet whether the proposed circuit worked.
4. Now, use the materials to create two of three different types of circuits.
5. Draw two of these circuits you created on your experiment sheet and write down what you have learned.

Vocabulary & Memory Work

- Vocabulary: parallel circuit, resistance, series circuit
- Memory Work—This week, work on memorizing the Ohm's Law.
  - Voltage (V) = Current (I) • Resistance (R)

Sketch: Circuit Diagram

- Label the following – Battery, resistor, switch, light
- Draw a circuit using the symbols on the sketch sheet that depict a pair of bulbs in series and a set of parallel-connected pair of bulbs.

Writing

- Reading Assignment: *DK Encyclopedia of Science* pp. 152-153 (Circuits)
- Additional Research Readings
  - Electrical Circuits: *KSE* pp. 340-341
  - Resistance: *KSE* pg. 363
  - Controlling Current: *UIDS* pp. 62-63

Dates

- 1827 – George Ohm publishes his work on resistance, including an equation that eventually becomes known as Ohm's Law.
- 1828 – Andre Ampere is elected as a member of the Royal Swedish Academy of Science in recognition of his contributions to the creation of the field of modern electrical science.

Sketch Week 24

Symbols

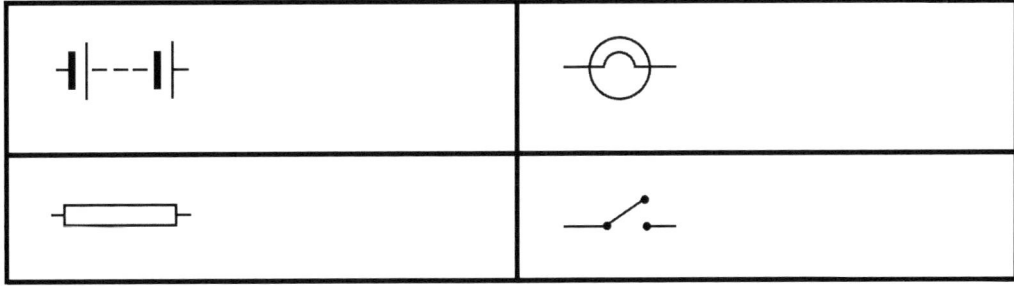

Online Circuits Lab

Hypothesis (Draw a circuit that you think will work.**)**

Did your proposed circuit work?   yes   no

My Circuits (Draw two of the other working circuits you designed.**)**

Conclusion

_____

_____

_____

_____

_____

Written Assignment Week 24

Discussion Questions

1. What is a circuit?
2. How does a flashlight represent a simple circuit?
3. How does resistance affect the flow of current in a circuit?
4. How is resistance affected when the resistors are connected in series? In parallel?

_____

_____

_____

_____

_____

_____

_____

_____

_____

_____

_____

_____

_____

_____

_____

_____

_____

_____

_____

_____

_____

_____

Written Assignment Week 24

Student Assignment Sheet Week 25
Magnetism

Experiment: Do different types of magnets have different strengths?

Materials
- ✓ 2 different types of magnets, such as a horseshoe magnet and a neodymium magnet
- ✓ Paper clips (20 to 30)
- ✓ Paper
- ✓ Cardboard
- ✓ Thick books

Procedure

1. Read the introduction to the experiment and answer the question for the hypothesis section.

2. Use the following tests to determine the strength of each of your magnets:
   - **Paper clip test** – Pick up one paper clip with the magnet. Then attach another one to the bottom of the first, creating a chain of paper clips. Continue to add more paper clips until the chain breaks or you cannot add anymore. Count how many paper clips you added and record that on your experiment sheet.
   - **Thickness test** – Have a partner hold a piece of paper with a paper clip on it. Use the magnet to move the paper clip around. Next, switch the paper for a piece of cardboard and repeat. Then, switch the cardboard for a thick book. Keep adding books until the magnet can no longer attract the paper clip and move it. Measure the thickness of the material that the magnet could attract through and record that on your experiment sheet.

3. Draw conclusions and complete the experiment sheet.

Vocabulary & Memory Work

- ☐ Vocabulary: electrolyte, magnetic field, magnetic pole
- ☐ Memory Work—This week, work on memorizing the first law of magnetism.
  - ↳ **First law of magnetism** – Like poles repel, while unlike poles attract.

Sketch: Magnetic Attraction and Repulsion

- 🖾 Draw the magnetic fields that show the repulsive force on one set of magnets and the attractive force on the other set.
- 🖾 Label the following – south pole, north pole, repulsive forces, attractive forces

Writing

- ᘓ Reading Assignment: *DK Encyclopedia of Science* pp. 154-155 (Magnetism)
- ᘓ Additional Research Readings
  - 📖 Magnets and Magnetism: *KSE* pp. 342-343
  - 📖 Magnets: *UIDS* pp. 70-71
  - 📖 Magnetic Fields: *UIDS* pp. 72-73

Dates

- 🕘 1100 – Chinese sailors are the first on record for using a magnetic compass for navigating on a cloudy day.
- 🕘 1600 – William Gilbert, an English doctor and physicist, publishes a book which for the first time explains exactly how a compass works.
- 🕘 1952 – Magnets are made out of ceramics for the first time.
- 🕘 1983 – Neodymium magnets are first invented.

Sketch Week 25

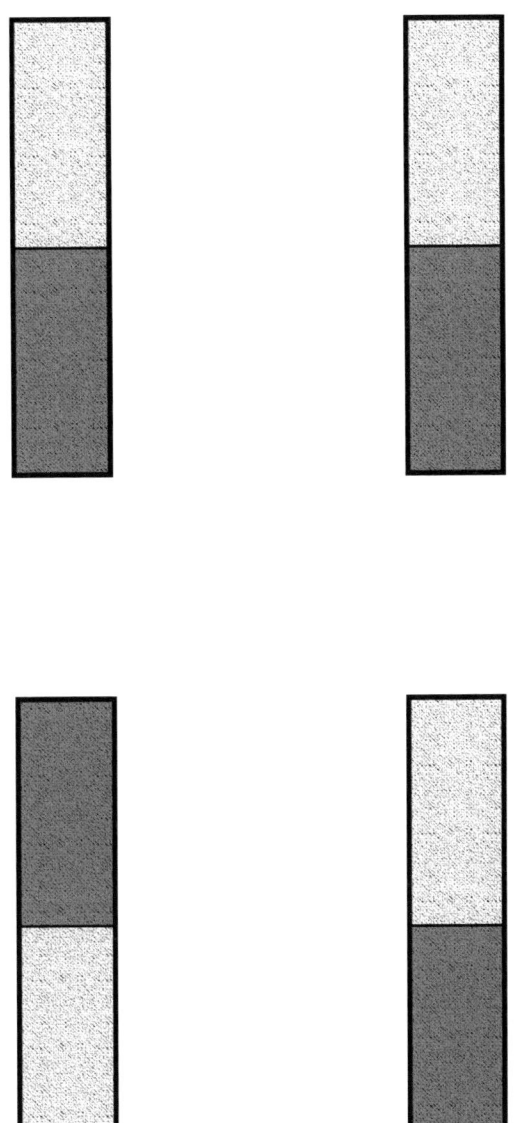

Do different types of magnets have different strengths?

Introduction

Magnets come in all shapes and sizes, like bars, horseshoes, discs, or rings. Three key metals have the ability to become permanently magnetized – cobalt, iron, and nickel. This property is known as ferromagnetism and it is the key to making magnets. In today's experiment, you are going to see if all magnets have the same strength or if their attractive force is dependent upon the materials of which they are composed.

Hypothesis

_____

_____

_____

Materials

_____   _____

_____   _____

_____   _____

Procedure

_____

_____

_____

_____

_____

_____

_____

_____

_____

_____

Observations and Results

	Paper Clip Test	Thickness Test
Magnet #1		
Magnet #2		

_____

_____

_____

Conclusion

_____

_____

_____

_____

_____

Written Assignment Week 25

Discussion Questions

1. What do all magnets have?
2. What causes the Earth's magnetic field? How does the magnetic field affect life on the surface of the Earth?
3. What causes a piece of steel to magnetize?
4. What can happen with you hit a steel magnet with a hammer?
5. What is the purpose of a keeper on a horseshoe magnet?

Written Assignment Week 25

Student Assignment Sheet Week 26
Electromagnetism

Experiment: Can I create a magnetic field with electricity?

Materials
- ✓ D battery
- ✓ Insulated copper wire – about 3 ft (1 m)
- ✓ 2 to 3 in (5 to 8 cm) Nail
- ✓ Electrical tape
- ✓ Iron filings
- ✓ Paper

Procedure
1. Read the introduction to the experiment and answer the question for the hypothesis section.
2. Wrap the nail with the insulated wire in tight coils, leave at least a 6 in (16 cm) tail on each end. Using the tape, attach one end of the wire to the positive terminal on the battery and the other end to the negative terminal. (*See the diagram above to visually check your set-up.*)
3. Sprinkle some of the iron filings on a piece of paper. Take hold of the battery in the center and move the wire-wrapped nail close to the filings and observe what happens. If you see a magnetic field form in the iron filings, draw the field on your experiment sheet.
4. Draw conclusions and complete the experiment sheet.

Vocabulary & Memory Work
- ☐ Vocabulary: electromagnet, electromagnetic field
- ☐ Memory Work—This week, continue to work on memorizing the first law of magnetism.

Sketch: Magnetic Fields in an Electromagnet
- ▨ Draw several arrows show the flow of the electrical current in the circuit and the magnetic field created by the electrical current.
- ▨ Label the following – south pole, north pole

Writing
- ☞ Reading Assignment: *DK Encyclopedia of Science* pp. pp. 156-157 (Electromagnetism)
- ☞ Additional Research Readings
  - 📖 Electromagnetism: *KSE* pp. 344-345, *UIDS* pp. 74-75

Dates
- 🕐 1820 – Hans Christiaan Oersted is performing an experiment with electrical current when he noticed that the needle of a nearby compass moved. This realization led to the discovery of electromagnetism.

Sketch Week 26

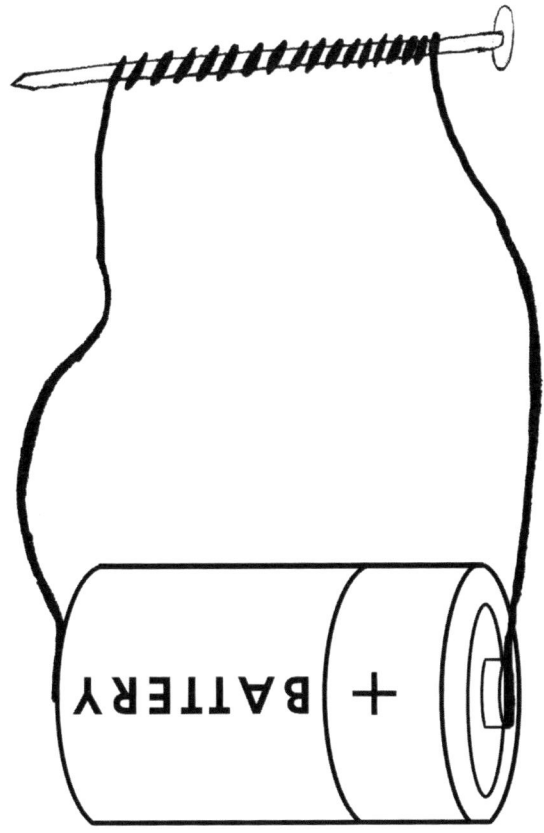

Can I create a magnetic field with electricity?

Introduction

All magnets have a magnetic field in which the force of the magnetic attraction can be felt. If we place iron filings near a magnet, the filings will line up in a pattern that shows the field produced by the magnet. In today's experiment, you are going to test to see if you can use electricity to yield the same effect.

Hypothesis

_____

_____

_____

Materials

_____   _____

_____   _____

_____   _____

Procedure

_____

_____

_____

_____

_____

_____

_____

_____

_____

_____

_____

_____

Observations and Results

Draw what you observed when you brought your electromagnet close to the iron filings.

_____

_____

_____

Conclusion

_____

_____

_____

_____

Written Assignment Week 26

Discussion Questions

1. What does an electrical current always produce?
2. What are two benefits of using electromagnets over permanent magnets?
3. How does a magnetic levitation train work?
4. What happens when electricity flows through a coil of wires?
5. What are some items that we encounter in everyday life that use electromagnetism?

Written Assignment Week 26

Student Assignment Sheet Week 27
Motors and Generators

Experiment: Can I raise a needle from a table without touching it?

Materials
- ✓ Straws
- ✓ Electrical tape
- ✓ 6 ft. of thin insulated wire
- ✓ AAA battery
- ✓ Sandpaper
- ✓ Needle
- ✓ *Robotics* book

Procedure
1. Read the introduction to the experiment on pg. 50 of *Robotics*.
2. Follow the directions on pp. 50-51 of *Robotics* to build your solenoid. Then, use the solenoid to attempt to raise a needle from a table without touching it.
3. Draw conclusions and complete the experiment sheet.

Vocabulary & Memory Work
- ☐ Vocabulary: power, solenoid, turbine
- ☐ Memory Work—There is no memory work this week.

Sketch: Fleming's Hand-rules
- ☒ Label the following – Left-hand rule for motors, thumb gives the wire's direction of motion, first finger shows the magnetic field's direction, second finger shows the current's direction, right-hand rule for generators, thumb give the direction of motion, first finger shows the magnetic field's direction, second finger shows the current's direction.

Writing
- ✍ Reading Assignment: *DK Encyclopedia of Science* pg. 158 (Electric Motors), pg. 159 (Generators)
- ✍ Additional Research Readings
    - 📖 Generators and Motors: *KSE* pp. 346-347
    - 📖 Electromagnetism (the motor effect): *UIDS* pg. 76

Dates
- 🕐 1821 – Michael Faraday discovers that electricity can produce rotary motion.

Sketch Week 27

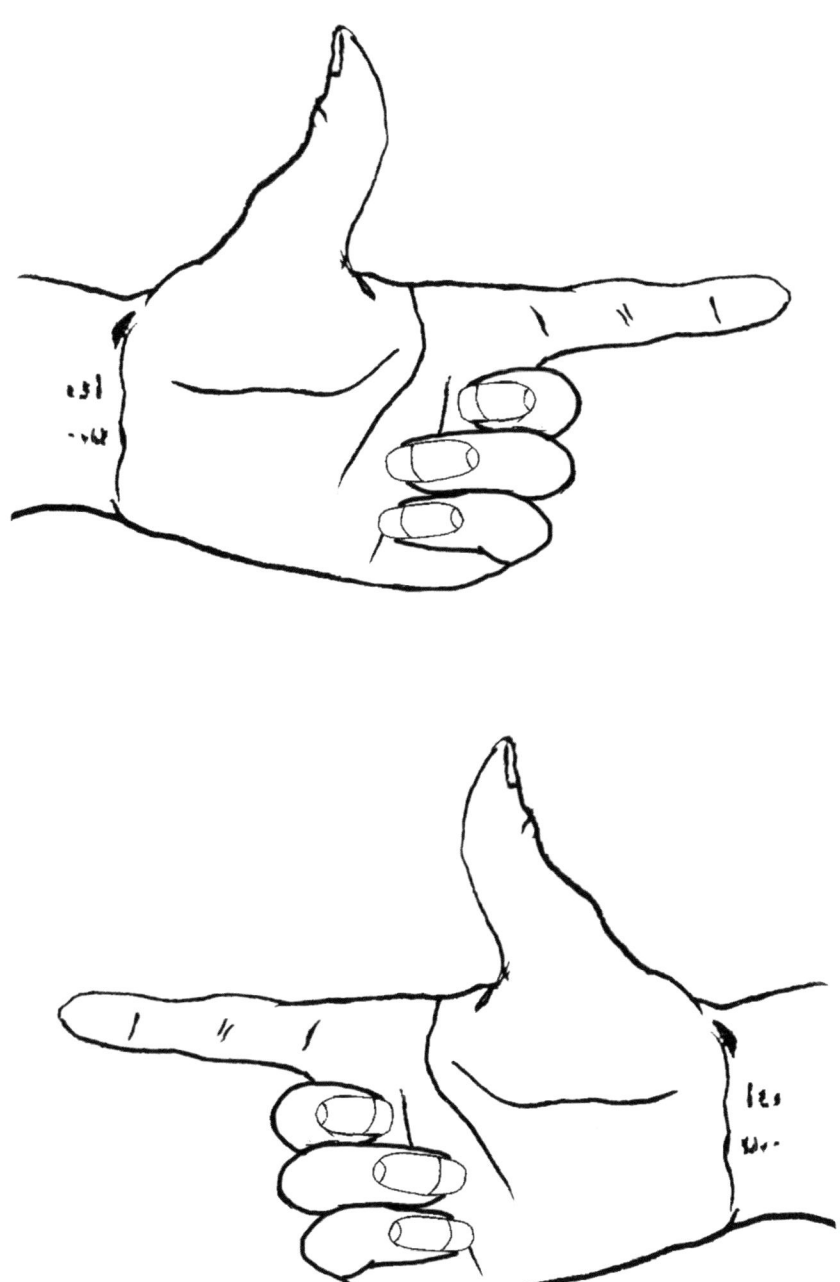

Can I raise a needle from a table without touching it?

Introduction

Read the introduction about solenoids found on pg. 50 of *Robotics*.

Hypothesis

_____

_____

_____

Materials

_____   _____

_____   _____

_____   _____

Procedure

_____

_____

_____

_____

_____

_____

_____

_____

_____

_____

_____

Observations and Results

_____

_____

_____

_____

_____

_____

_____

_____

_____

_____

Conclusion

_____

_____

_____

_____

_____

Written Assignment Week 27

Discussion Questions

Motors, pg. 158

    1. What does an electric motor do?

    2. How does an electric motor work?

    3. Why are electric motors good sources of power?

Generators, pg. 159

    1. How do generators differ from motors?

    2. What is electromagnetic induction?

Written Assignment Week 27

Physics
Unit 7

Engineering and Robotics

Unit 7 Engineering and Robotics
Vocabulary Sheet

Define the following terms as they are assigned on your Student Assignment Sheet.

1. Engineer – _____

_____

_____

2. Load – _____

_____

_____

3. Beam – _____

_____

_____

4. Bridge – _____

_____

_____

5. Truss – _____

_____

_____

6. Arch – _____

_____

_____

7. Tunnel – _____

_____

_____

8. Automaton – _____

_____

_____

9. Robot – _____

_____

_____

10. Actuator – _____

_____

_____

11. Capacitor – _____

_____

_____

12. Effector – _____

_____

_____

13. Controller – _____

_____

_____

14. Photoresistor – _____

_____

_____

15. Sensor – _____

_____

_____

Student Assignment Sheet Week 28
Engineering

Experiment: Which shape is stronger?
Materials
- ✓ Paper
- ✓ Tape
- ✓ Books
- ✓ Can or glass

Procedure
1. Read the introduction to the experiment on pg. 22 of *Bridges and Tunnels*.
2. Follow the directions on pg. 22 of *Bridges and Tunnels* to make your shapes. Then, test which one of the shapes you made is stronger.
3. Draw conclusions and complete the experiment sheet.

Vocabulary & Memory Work
- ☐ Vocabulary: engineer, load
- ☐ Memory Work—This week, work on memorizing the engineering design process.
  - ↓ **The Engineering Design Process**
    1. Identify the Need or Problem
    2. Research and Define Requirements
    3. Brainstorm for Solutions
    4. Design a Plan
    5. Build a Prototype
    6. Test and Evaluate the Prototype
    7. Communicate the Results
    8. Redesign if Needed

Sketch: 5 Main Branches of Engineering
- 🖼 Use the information found in the chart on pg. 9 of *Bridges and Tunnels* to fill in the chart.

Writing
- ✍ Reading Assignment: *Bridges and Tunnels* pp. 1-6 Lifelines, pp. 7-17 Engineering and Thinking Big
- ✍ Additional Research Readings
  - 📖 Brick, Stone, and Concrete: *KSE* pp. 220-221

Dates
- 🕐 520 BC – Greek architect, Eupalinos, becomes the first engineer.
- 🕐 27 BC-393 AD – The Roman Empire constructs many arch bridges, some of which still stand today.

Sketch Week 28

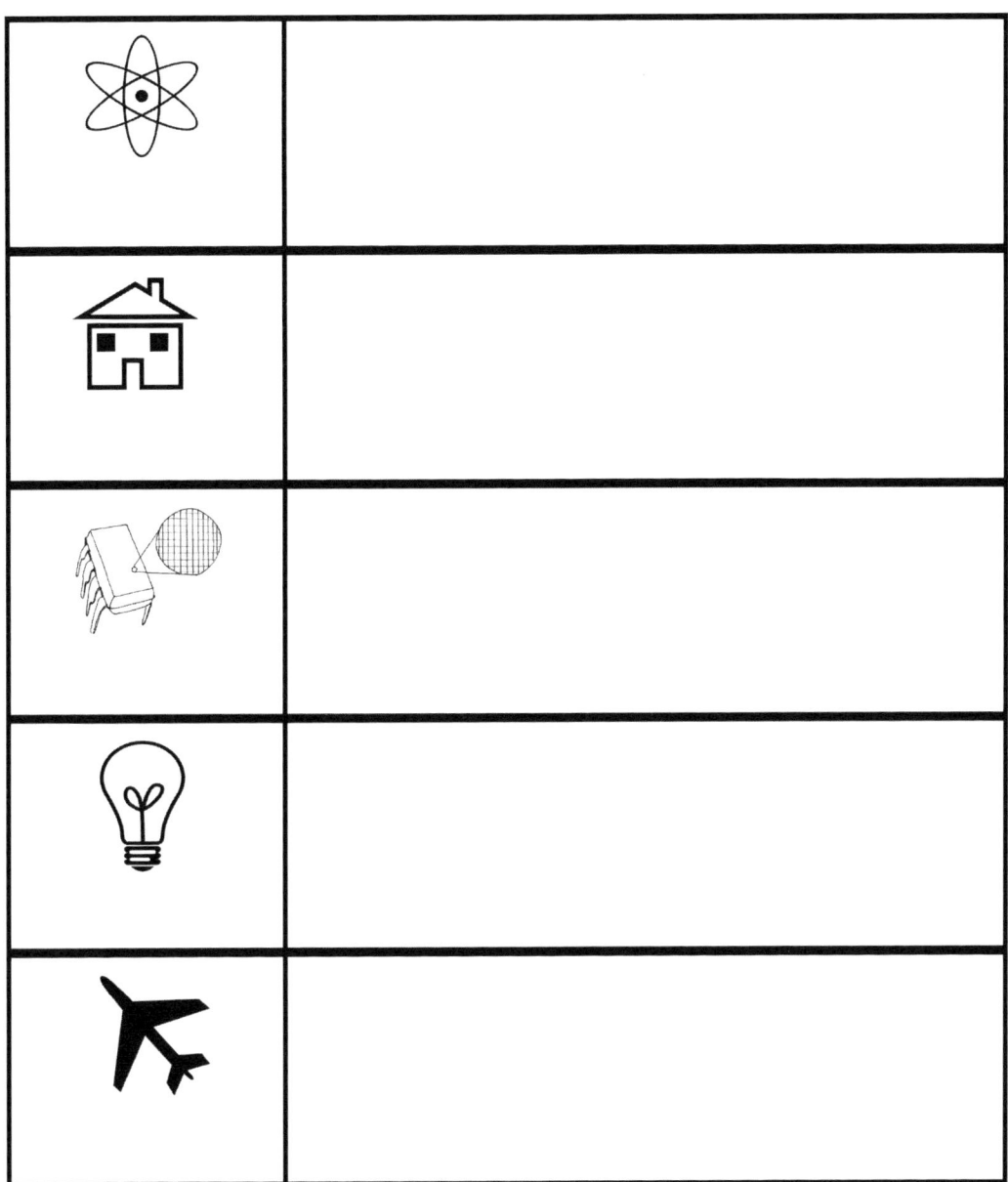

Which shape is stronger?

Introduction

Read the introduction about engineers and shapes found on pg. 22 of *Bridges and Tunnels*.

Hypothesis

_____

_____

_____

Materials

_____   _____

_____   _____

_____   _____

_____   _____

_____   _____

Procedure

_____

_____

_____

_____

_____

_____

_____

_____

_____

_____

_____

Observations and Results

Type of shape			
Amount of weight it held			

_____

_____

_____

_____

_____

_____

_____

Conclusion

_____

_____

_____

_____

Written Assignment Week 28

Discussion Questions

Lifelines, pp. 1-6

1. What do bridges and tunnels do?
2. What is trial-and-error?

Engineering and Thinking Big, pp. 7-17

1. What are the different branches of engineering?
2. What is tension? Compression?
3. How do tension and compression affect a structure?
4. What is torsion? Shear?
5. What natural phenomena can produce torsion and shear within a structure?

_____

_____

_____

_____

_____

_____

_____

_____

_____

_____

_____

_____

_____

_____

_____

_____

_____

_____

_____

_____

_____

Written Assignment Week 28

Student Assignment Sheet Week 29
Bridges

Experiment: How strong is a craft stick beam bridge?

   Materials
   - ✓ Craft sticks
   - ✓ Wood glue
   - ✓ Books
   - ✓ Binder clips

   Procedure
   1. Read the introduction to the experiment on pg. 58 of *Bridges and Tunnels*.
   2. Follow the directions on pg. 58-59 of *Bridges and Tunnels* to build your craft stick beam bridge. Then, test how strong the bridge is.
   3. Draw conclusions and complete the experiment sheet.

Vocabulary & Memory Work

   - ☐ Vocabulary: beam, bridge, truss
   - ☐ Memory Work—This week, continue to work on memorizing the engineering design process.

Sketch: 3 Types of Bridges

   - 🔲 Label the following – beam bridge, arch bridge, suspension bridge

Writing

   - ৶ Reading Assignment: *Bridges and Tunnels* pp. 23-31 Building Big: The Physics of Bridges, pp. 39-56 Amazing Bridges
   - ৶ Additional Research Readings
      - 📖 Construction: *KSE* pp. 223-223
      - 📖 When Bridges Collapse: *Bridges and Tunnels* Chapter 4

Dates

   - 🕐 62 BC – The Ponte Fabrico Bridge, a stone double-arch bridge, is built by the Roman Empire.
   - 🕐 1848 – One of the earliest wire suspension bridges is designed by Charles Ellet and built to span the Niagara River near Niagara Falls.
   - 🕐 1949 – The world's first geodesic dome is built by Richard Fuller.

Sketch Week 29

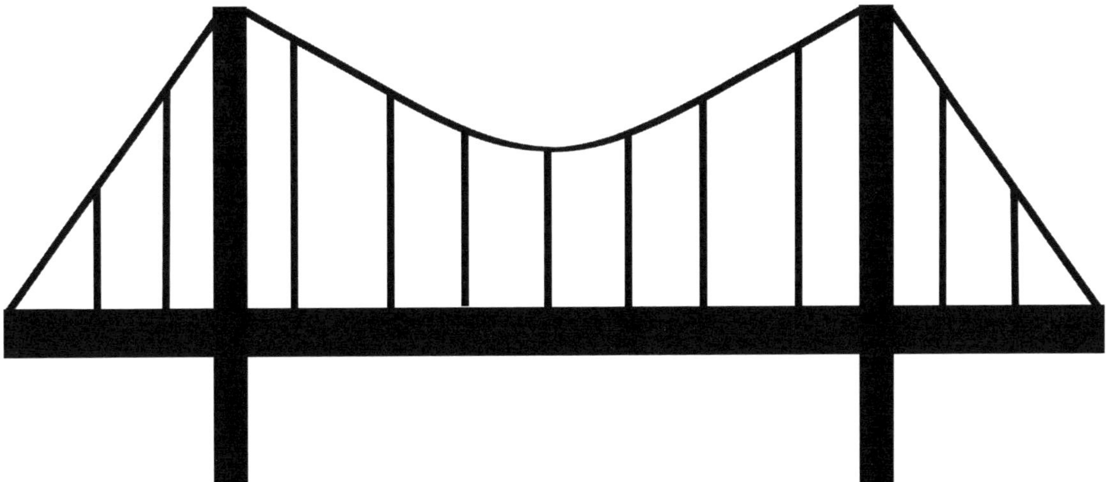

Student Guide Unit 7 Engineering and Robotics ~ Week 29 Bridges

212

How strong is a craft stick beam bridge?

Introduction

Read the introduction about beam bridges found on pg. 58 of *Bridges and Tunnels*.

Hypothesis

_____

_____

Materials

_____   _____

_____   _____

_____   _____

Procedure

_____

_____

_____

_____

_____

_____

_____

_____

_____

_____

_____

_____

Observations and Results

_____

_____

_____

_____

_____

_____

_____

My Bridge

Conclusion

_____

_____

_____

_____

_____

_____

_____

Written Assignment Week 29

Discussion Questions

Building Big: The Physics of Bridges, pp. 23-31
1. What are the three major bridge designs?
2. How is a beam bridge constructed and when is it useful?
3. How is an arch bridge constructed and when is it useful?
4. How is a suspension bridge constructed and when is it useful?

Amazing Bridges, pp. 39-56
1. What are caissons and cofferdams?
2. What does the keystone do in an arch?
3. Tell the story of how the Brooklyn Bridge or Golden Gate Bridge was built.

_____

_____

_____

_____

_____

_____

_____

_____

_____

_____

_____

_____

_____

_____

_____

_____

_____

_____

_____

_____

Written Assignment Week 29

Student Assignment Sheet Week 30
Tunnels

Experiment: Tunnel Building Project
Materials
- ✓ Salt dough (at least 3 to 4 cups)
- ✓ Spoon
- ✓ Craft sticks
- ✓ Pipe cleaners
- ✓ Aluminum foil
- ✓ Cardboard square
- ✓ Toy car
- ✓ Books or other heavy objects
- ✓ Water

Procedure
1. Read the introduction to the experiment.
2. Set the salt dough on the cardboard square and shape the dough into a mountain.
3. Then, use the spoon to dig a tunnel at the base of the mountain large enough for the toy car to drive through. Add supports, pipe cleaners, and craft sticks, and a foil lining for the tunnel as you dig.
4. Once the tunnel is complete, set the mountain aside and let it dry for several days.
5. After the mountain is dry, head outside and place the toy car in the tunnel. Then, slowly pour water over your mountain and check to see if any water seeps into your tunnel.
6. Next, use the books or other heavy objects to test the whether your tunnel will collapse.
7. Draw conclusions and complete the experiment sheet.

Vocabulary & Memory Work
- ☐ Vocabulary: arch, tunnel
- ☐ Memory Work—This week, continue to work on memorizing the engineering design process.

Sketch: Forces in a Tunnel
- ▣ Draw and label the forces that are present in a tunnel – compression forces from the weight above, pushing forces from the solid rock weight below

Writing
- ✐ Reading Assignment: *Bridges and Tunnels* pp. 74-84 Building Big: The Physics of Tunnels
- ✐ Additional Research Readings
  - 📖 Amazing Tunnels: *Bridges and Tunnels* Chapter 6
  - 📖 Tunnel Disasters: *Bridges and Tunnels* Chapter 7

Dates
- 🕐 1869-1883 – The Brooklyn Bridge, a steel-cable suspension bridge, is built.
- 🕐 1899 – Alfred Nobel invents dynamite, which changes tunnel and bridge building forever.

Sketch Week 30

Tunnel Building Project

Introduction

Tunnels are built to get through mountains and other barriers. Engineers have to design supports that prevent the tunnel from collapsing and keep water from filling the tunnel. In today's experiment, you will be engineering your own tunnel through a salt dough mountain.

Hypothesis

_____

_____

Materials

_____   _____

_____   _____

_____   _____

Procedure

_____

_____

_____

_____

_____

_____

_____

_____

_____

_____

Observations and Results

_____

_____

_____

_____

_____

_____

_____

My Tunnel

Conclusion

_____

_____

_____

_____

_____

_____

_____

Written Assignment Week 30

Discussion Questions

1. What are two ways that natural tunnels can be formed?
2. What are some ways that humans have used tunnels?
3. What are the three stages of man-made tunnel building? Explain what happens in each.
4. What are the three tunnel types?

Written Assignment Week 30

Student Assignment Sheet Week 31
Robotics

Experiment: Vibrorobot
    Materials
- ✓ 1.5-volt DC motor
- ✓ 1 ft. insulated wire
- ✓ Electrical tape
- ✓ Cup or Jar
- ✓ Foam tape
- ✓ 2 AAA batteries
- ✓ Rubber band
- ✓ Cork
- ✓ Cardboard
- ✓ 3 Pens
- ✓ Paper

    Procedure
1. Read the introduction to the experiment on pg. 24 of *Robotics*.
2. Follow the directions on pg. 24-26 of *Robotics* to build a vibrating robot. Then, test how the robot works as it draws a picture for you.
3. Draw conclusions and complete the experiment sheet.

Vocabulary & Memory Work
- ☐ Vocabulary: automata, robot
- ☐ Memory Work—There is no memory work for this week.

Sketch: Sense – Think – Act Cycle
- ▨ Label the following – the robot takes in information, the robot uses the information to choose the next step, the robot does something.

Writing
- ᗯ Reading Assignment: *Robotics* pp. 11-23 Development of Robotics, pp. 27-33 Housing: Robot Bodies
- ᗯ Additional Research Readings
  - 📖 Introduction: *Robotics* pp. 1-7

Dates
- 🕐 200 BC – The Chinese create mechanical musicians to play music for the emperor.
- 🕐 1555 – Gianello Torriano, an Italian clock maker, makes a model of a wind-up lady that can walk in a circle and strum a lute.
- 🕐 1822 – Charles Babbage designs a mechanical calculator that uses punch cards.
- 🕐 1959 – The Massachusetts Institute of Technology (MIT) opens the first lab to study artificial intelligence.

Sketch Week 31

Sense

Think

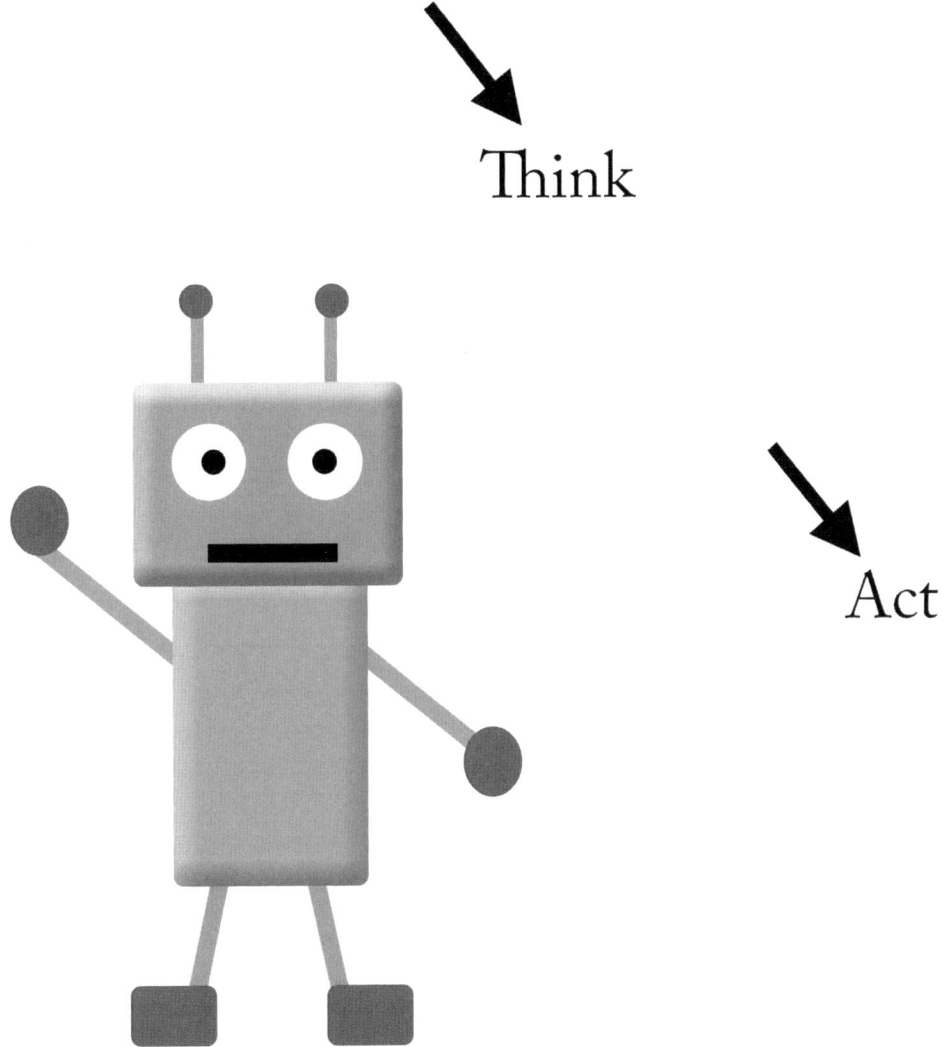

Act

Vibrorobot

Introduction

Read the introduction on vibrorobots found on pg. 24 of *Robotics*.

Hypothesis

_____

_____

Materials

_____   _____

_____   _____

_____   _____

Procedure

_____

_____

_____

_____

_____

_____

_____

_____

_____

_____

_____

_____

Observations and Results

_____

_____

_____

_____
_____
_____
_____

My Vibrorobot

Conclusion

_____
_____
_____
_____
_____
_____

Written Assignment Week 31

Discussion Questions

Development of Robots, pp. 11-23

    1. What kinds of automata existed before modern-day robots?

    2. What do people use robots for?

Housing: Robot Bodies, pp. 27-33

    1. What are nanobots?

    2. What is a robotic swarm?

    3. What is a biomimetic robot?

Written Assignment Week 31

Student Assignment Sheet Week 32
Actuators and Effectors

Experiment: Wobblebot

Materials
- ✓ Pencil
- ✓ 1.5-volt DC motor
- ✓ Small Solar Panel
- ✓ Electrical tape
- ✓ Scissors
- ✓ CD
- ✓ Glue
- ✓ Tape
- ✓ Clear dome from a drink cup

Procedure
1. Read the introduction to the experiment on pg. 48 of *Robotics*.
2. Follow the directions on pg. 48-49 of *Robotics* to build a solar wobblebot. Then, test how the robot works.
3. Draw conclusions and complete the experiment sheet.

Vocabulary & Memory Work
- ☐ Vocabulary: actuator, capacitor, effector
- ☐ Memory Work—There is no memory work for this week.

Sketch: Anatomy of a Solar Cell
- ▨ Label the following – solar cell, junction, negative panel, positive panel, the sun knocks electrons off atoms and sets them in motion, the electrons are used to power a device.

Writing
- ✑ Reading Assignment: *Robotics* pp. 38-47 Actuators: Making Robots Move, pp. 55-59 Effectors: How Robots Do Things
- ✑ Additional Research Readings
  - 📖 Robotics: *KSE* pp. 236-237
  - 📖 Robots: *DK EOS* pg. 176

Dates
- ⏲ 1989 – Mark Tilden coins the term "BEAM robot" to refer to a simple solar-powered, life-like robot.
- ⏲ 2000 – The Honda car company develops ASIMO, which revolutionized the way robots get around.

Sketch Week 32

Wobblebot

Introduction

Read the introduction on BEAM robots found on pg. 48 of *Robotics*.

Hypothesis

_____

_____

Materials

_____   _____

_____   _____

_____   _____

Procedure

_____

_____

_____

_____

_____

_____

_____

_____

_____

_____

_____

_____

_____

Observations and Results

_____

_____

_____
_____
_____
_____

My Wobblebot

Conclusion

_____
_____
_____
_____
_____
_____
_____

Written Assignment Week 32

Discussion Questions

Actuators: Making Robots Move, pp. 38-47

1. What does BEAM stand for?
2. How does the addition of a capacitor help a BEAM robot?
3. What are two ways, other than a battery, to power a robot?
4. What type of actuators can be used in a robot?

Effectors: How Robots Do Things, pp. 55-59

1. What are several examples of effectors?
2. What does "degree of freedom" refer to?
3. What does each degree of freedom need?

_____

_____

_____

_____

_____

_____

_____

_____

_____

_____

_____

_____

_____

_____

_____

_____

_____

_____

_____

_____

_____

Written Assignment Week 32

Student Assignment Sheet Week 33
Sensors and Controllers

Experiment: Pressure Sensor
  Materials
  - ✓ LED light bulb with two metal legs
  - ✓ 3-volt Watch battery
  - ✓ 2 Index cards
  - ✓ Aluminum foil
  - ✓ Scissors
  - ✓ Marker
  - ✓ Yarn
  - ✓ Glue
  - ✓ Toothpick
  - ✓ Tissue

  Procedure
  1. Read the introduction to the experiment on pg. 80 of *Robotics*.
  2. Follow the directions on pg. 80-81 of *Robotics* to build a pressure sensor. Then, test how the sensor works.
  3. Draw conclusions and complete the experiment sheet.

Vocabulary & Memory Work
  - ☐ Vocabulary: controller, photoresistor, sensor
  - ☐ Memory Work—There is no memory work for this week.

Sketch: Anatomy of a Light Sensor
  - ☒ Label the following – electrodes, photoconductive material, the sensor receives light input, the sensor sends information to processor.

Writing
  - ∾ Reading Assignment: *Robotics* pp. 68-78 Sensors: How Robots Know What's Going On, pp. 85-93 Controller: How Robots Think
  - ∾ Additional Research Readings
    - 📖 Social Robots: *Robotics* pp. 104-113

Dates
  - 🕐 1947 – The transistor is invented by John Bardeen, Walter Brattain, and William Shockley. This invention makes it possible to design modern computers and robots.
  - 🕐 1967 – Seymor Papert, an MIT mathematician, develops Logo, one of the earliest computer languages for children.

Sketch Week 33

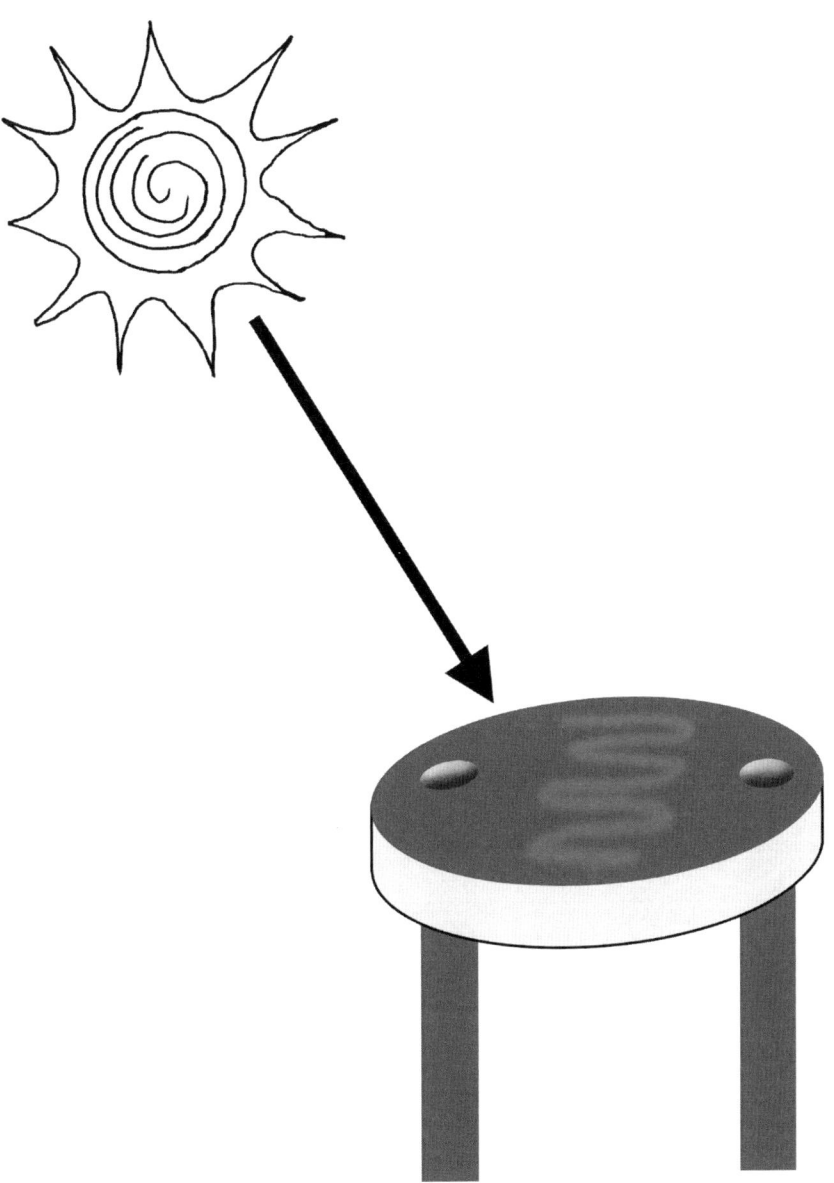

Pressure Sensor

Introduction

Read the introduction on pressure sensors found on pg. 80 of *Robotics*.

Hypothesis

_____

_____

Materials

_____   _____

_____   _____

_____   _____

Procedure

_____

_____

_____

_____

_____

_____

_____

_____

_____

_____

_____

_____

Observations and Results

_____

_____

_____

————————————————————————

————————————————————————

————————————————————————

————————————————————————

My Pressure Sensor

Conclusion

————————————————————————

————————————————————————

————————————————————————

————————————————————————

————————————————————————

————————————————————————

————————————————————————

Written Assignment Week 33

Discussion Questions

Sensors: How Robots Know What's Going On, pp. 68-78

1. How do a robot's sensors work?
2. What different types of sensors can robots have?

Controller: How Robots Think, pp. 85-93

1. What is a micro-controller?
2. How are computers and code used in robotics?
3. What is the most important part of computer programming?
4. What is Boolean logic?

_____

_____

_____

_____

_____

_____

_____

_____

_____

_____

_____

_____

_____

_____

_____

_____

_____

_____

_____

_____

_____

_____

Written Assignment Week 33

Physics
Unit 8
Nuclear Physics

242

Unit 8 Nuclear Physics
Vocabulary Sheet

Define the following terms as they are assigned on your Student Assignment Sheet.

1. Radiation – _____

   _____

   _____

2. Radioactive decay – _____

   _____

   _____

3. Radioisotope – _____

   _____

   _____

4. Fission – _____

   _____

   _____

5. Fusion – _____

   _____

   _____

6. Nuclear reactor – _____

   _____

   _____

Student Guide Unit 8 Nuclear Physics ~ Vocabulary

Student Assignment Sheet Week 34
Radioactivity

Scientist Biography Report Project

This week, you will complete step one and two of your Scientist Biography Report Project. You will be choosing the scientist you would like to learn more about and do your research. The instructions for this week's assignments are on the following Scientist Biography Report sheets.

Vocabulary & Memory Work

- ☐ Vocabulary: radiation, radioactive decay, radioisotope
- ☐ Memory Work—This week, begin to work on memorizing the types of radioactive particles.
  - ↯ **Types of Radioactive Particles**
    1. **Alpha Particles** – Positively charged particles that can be ejected from some radioactive nuclei. Each of the particles consist of two protons and two neutrons.
    2. **Beta Particles** – Particles that are ejected from a radioactive nucleus at the speed of light. Each particle has the same mass as an electron.
    3. **Gamma Rays** – Electromagnetic waves that have high penetrating power and are generally emitted from a radioactive nucleus after alpha and beta particles.

Sketch: Radioactive Penetrating Power

- ▨ Label the following – alpha particle, beta particle, gamma rays, thick sheet of paper, sheet of aluminum, thick sheet of lead
- ▨ Draw the arrows show the penetrating abilities of the rays.

Writing

- ⌇ Reading Assignment: *DK Encyclopedia of Science* pp. 26-27 Radioactivity
- ⌇ Additional Research Readings
  - 📖 Radiation: *KSE* pp. 244-245
  - 📖 Radioactivity: *UIDS* pp. 86-87
  - 📖 Uses of Radioactivity: *UIDS* pg. 91

Dates

- 🕐 1896 – Antoine Becquerel is working with a natural fluorescent material and x-rays, which leads to his discovery of radioactivity.
- 🕐 1898 – Marie and Pierre Curie coin the term "radioactive" when they discovered radium and polonium.
- 🕐 1928 – Hans Geiger and one of his students develop a machine that can detect and measure the intensity of radiation.

Sketch Week 34

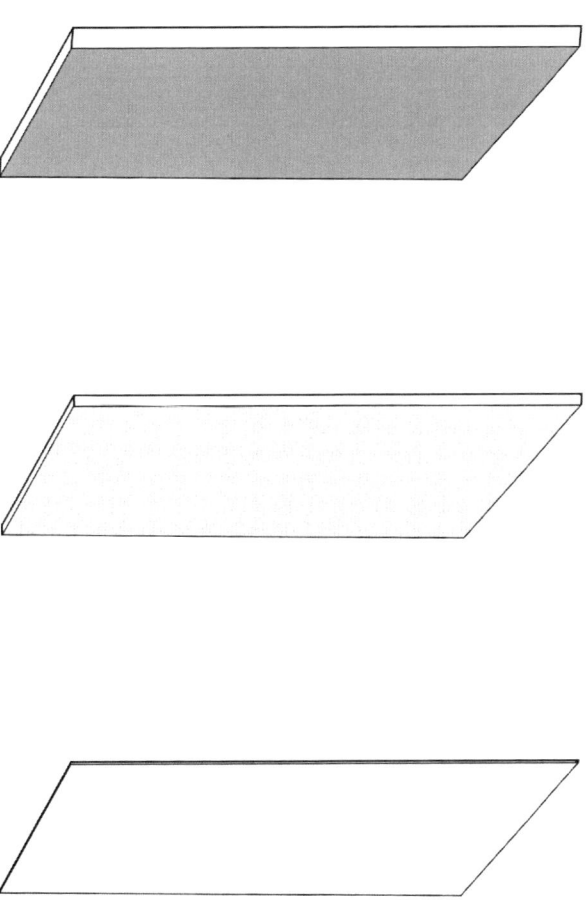

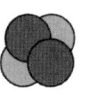 

Scientist Biography Report

Step 1: Choose a Scientist

During the next three weeks, you are going to be researching and learning more about a scientist that has contributed to the field of physics. This week, you need to begin your scientist biography project by choosing which scientist you will research. You can choose one of the scientists mentioned in the "Dates" sections or you can choose one that has interested you.

The scientist I am going to study is:

_____

Step 2: Research the Scientist

Once you have chosen the scientist you would like to study, you can begin your research. Begin by looking for a biography on your chosen scientist at the library. Then, look for articles on the chemist in magazines, newspapers, encyclopedias, or on the Internet. You will need to know the following about your scientist to write your report:

☑ **Biographical information on the scientist** (*i.e., where they were born, their parents, siblings, and how they grew up*);

☑ **The scientist's education** (*i.e., where they went to school, what kind of student they were, what they studied, and so on*);

☑ **Their scientific contributions** (*i.e., research that they participated in, any significant discoveries they made, and the state of the world at the time of their contributions*).

As you read over the material you have gathered, be sure to write down any facts you glean in your own words. You can do this on the sheet below or on separate index cards.

My Research

_____

_____

_____

_____

_____

_____

_____

_____

_____

_____

Written Assignment Week 34

Discussion Questions

1. What is meant by a radioactive atom?
2. How is the number of subatomic particles related to an atom's potential radioactivity?
3. Why are radioactive materials often stored in water?
4. What are some helpful uses of radiation?
5. What is radioactive fallout?
6. What is a half-life?

_____

_____

_____

_____

_____

_____

_____

_____

_____

_____

_____

_____

_____

_____

_____

_____

_____

_____

_____

_____

Written Assignment Week 34

Student Assignment Sheet Week 35
Nuclear Energy

Scientist Biography Report Project

This week, you will complete step three through five of the Scientist Biography Report Project. You will be finishing up your research and organizing the information you have gathered into an outline. The instructions for this week's assignments are on the following Scientist Biography Report sheets.

Vocabulary & Memory Work

☐ Vocabulary: fission, fusion, nuclear reactor

☐ Memory Work—This week, continue to work on memorizing the types of radioactive particles.

Sketch: Nuclear Fission

▨ Label the following – high speed neutron, unstable nucleus, energy, neutrons, nuclei

Writing

✍ Reading Assignment: *DK Encyclopedia of Science* pp. 136-137 Nuclear Energy

✍ Additional Research Readings

📖 Power Plants (section on Nuclear Power): *KSE* pp. 348-349

📖 Atomic and Nuclear Energy: *UIDS* pp. 84-85

📖 Nuclear fission and fusion: *UIDS* pp. 92-93

Dates

🕐 1919 – Ernst Rutherford changes the nucleus of a nitrogen atom into an oxygen nucleus.

🕐 1956 – The first nuclear power plant, located in England, starts generating power.

🕐 1986 – The nuclear reactor at Chernobyl melts down, throwing radioactive material into the air.

Sketch Week 35

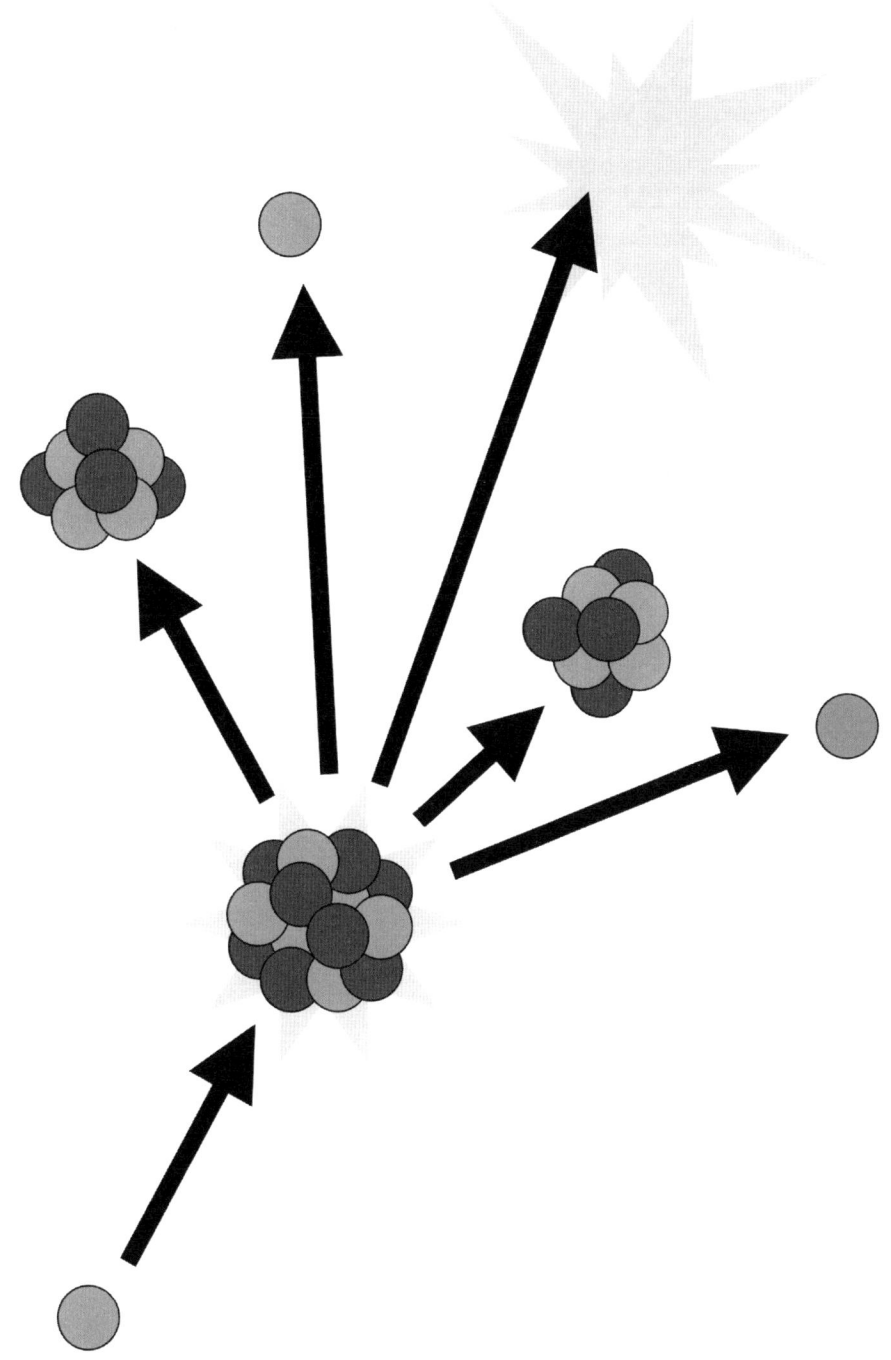

Step 3: Create an Outline

Now that your research is completed, you are ready to begin the process of writing a report on your chosen scientist. This week, you are going to organize the notes you took during step two into a formal outline which you will use next week to write the rough draft of your report. Use the outline template provided below as a guide. You should include information on why you chose the particular scientist in your introduction section. For the conclusion section of the outline, you need to include why you believe someone else should learn about your chosen scientist and your impression of the scientist (i.e., *Did you like the scientist? Do you feel that they made a significant impact on the field of physics?*).

Scientist Biography Outline

I. Introduction and Biological Information on the Scientist

    A. _____

    _____

    B. _____

    _____

    C. _____

    _____

    D. _____

    _____

    E. _____

    _____

    F. _____

    _____

II. The Scientist's Education

    A. _____

    _____

    B. _____

    _____

    C. _____

    _____

    D. _____

E. _____

F. _____

III. The Scientist's Contributions

    A. _____

        1. _____

        2. _____

        3. _____

        4. _____

    B. _____

        1. _____

        2. _____

        3. _____

        4. _____

    C. _____

1. _____

_____

2. _____

_____

3. _____

_____

4. _____

_____

IV. Conclusion

A. _____

_____

B. _____

_____

C. _____

_____

D. _____

_____

E. _____

_____

F. _____

_____

Scientist Biography Report: Final Draft

Written Assignment Week 35

Discussion Questions

1. What is the basic difference between nuclear fission and fusion?
2. What elements are typically used in nuclear fuel rods?
3. Why are boron rods found in a nuclear reactor?
4. What is the difference between an atomic bomb and a hydrogen bomb?
5. Why is fusion not a practical method of producing power on Earth yet?

Written Assignment Week 35

Appendix

Physics Memory Work

Unit 1

Newton's Three Laws of Motion

1. An object will not move unless a force, like a push or pull, moves it. Once it is moving, an object will not stop moving in a straight line unless it's forced to change.
2. The greater the force on an object, the greater the change in its motion. The greater the mass of an object, the greater the force needed to change its motion.
3. For every reaction, there is an equal but opposite reaction.

Unit 2

Types of Energy

1. **Mechanical energy** – The energy associated with the motion and position of an object. It is the sum of an object's kinetic and potential energy.
2. **Chemical energy** – The energy that is stored in chemical bonds. It is released when these bonds are broken in a chemical reaction.
3. **Thermal energy** – The energy that flows from one place to another due to changes in temperature. It is also known as heat.
4. **Electrical energy** – The energy associated with electrical charges.
5. **Electromagnetic energy** – The energy that travels through space in the form of waves, such as x-rays, light, and sound.
6. **Nuclear energy** – The energy stored in an atomic nucleus.

Unit 3

Laws of Thermodynamics

1. **Zeroth** – When two systems are in equilibrium with a third system, they are said to be in thermal equilibrium with each other.
2. **First** – Energy cannot be created or destroyed (also known as the Law of Conservation of Energy).
3. **Second** – Disorder (entropy) in the universe is always increasing.
4. **Third** – There is a theoretical point at which all molecular movement stops, which is known as absolute zero.

Unit 4

Types of Mechanical Waves

1. **Transverse Wave** – A wave that causes the medium to vibrate perpendicular to the direction in which the wave travels.
2. **Longitudinal Waves** – A wave that causes the medium to vibrate parallel to the direction in which the wave travels.
3. **Surface Waves** – A wave that travels along the surface separating two media.

Unit 5

Waves of the Electromagnetic Spectrum

1. **Radio waves** – These waves are typically found in radios, televisions, microwaves, and radars.
2. **Infrared rays** – These waves are used as a source of heat as well as to find areas of heat differences.
3. **Visible light** – These waves make up the spectrum of visible light we can see.
4. **Ultraviolet rays** – These waves can kill microorganisms, produce vitamin D in our skin, and help plants grow.
5. **X-rays** – These waves have very short wavelengths and are used to make pictures of what is inside a solid object.
6. **Gamma rays** – These waves have the shortest wavelengths in the electromagnetic spectrum. Gamma rays are used in medicine to kill cancer cells and to provide images of the brain.

Unit 6

Law of Electrostatics

Like charges repel each other and opposite charges attract each other.

Types of Electrical Current

1. **DC** – Direct Current is when the electrical charge only flows in one direction.
2. **AC** – Alternating Current is when the electrical charge regularly reverses the direction of its flow.

First Law of Magnetism

Like poles repel, while unlike poles attract.

Unit 7

The Engineering Design Process

1. Identify the Need or Problem
2. Research and Define Requirements
3. Brainstorm for Solutions
4. Design a Plan
5. Build a Prototype
6. Test and Evaluate the Prototype
7. Communicate the Results
8. Redesign if Needed

Unit 8

Types of Radioactive Particles

1. **Alpha Particles** – Positively charged particles that can be ejected from some radioactive nuclei. Each of the particles consist of two protons and two neutrons.
2. **Beta Particles** – Particles that are ejected from a radioactive nucleus at the speed of light. Each particle has the same mass as an electron.
3. **Gamma Rays** – Electromagnetic waves that have high penetrating power and are generally emitted from a radioactive nucleus after alpha and beta particles.

Physics Equations

Unit 1 Equations

🜂 Force Unit

1 Newton (N) = 1 kilogram (kg) • 1 meter (m) / second (s^2)

🜂 Motion Equation

$F = m \cdot A$

"F" stands for net force.
"m" stands for mass.
"A" stands for acceleration.

🜂 Speed Equation

$v = \dfrac{d}{t}$

"v" stands for average speed.
"d" stands for distance.
"t" stands for time.

🜂 Acceleration Equation

$A = \dfrac{v_f - v_i}{t}$

"A" stands for acceleration.
"v_f" stands for final speed.
"v_i" stands for initial speed.
"t" stands for time.

Unit 2 Equations

🜂 Work Equation
$W = F \cdot d$

"W" stands for work.
"F" stands for force.
"d" stands for distance.

🜂 Pressure Equation
$P = \dfrac{F}{a}$

"P" stands for pressure.
"F" stands for force.
"a" stands for area.

Unit 3 Equations

🜂 Celsius to Fahrenheit Equation

$^{\circ}F = 1.8 \cdot {}^{\circ}C + 32$

"°F" stands for the temperature in Fahrenheit.
"°C" stands for the temperature in Celsius.

🜂 Specific Heat Equation

$Q = m \cdot c \cdot \Delta T$

"Q" stands for the heat that is absorbed by a material.
"m" stands for mass.
"c" stands for specific heat.
"ΔT" stands for change in temperature.

🜂 Power Equation

$P = \dfrac{W}{t}$

"P" stands for power (measured in watts).
"W" stands for work.
"t" stands for time.

Unit 4 and 5 Equation

🜂 Speed of Waves Equation
$v = \lambda \cdot f$

"v" stands for speed.
"λ" stands for wavelength.
"f" stands for frequency.

Unit 6 Equations

🜂 Electrical Charge Equation
$Q = I \cdot t$

"Q" stands for electrical charge (measured in coulombs).
"I" stands for current (measured in amps).
"t" stands for time.

✦ Potential Difference Equation

$$V = \frac{E}{C}$$

"V" stands for potential difference or voltage (measured in volts).
"E" stands for energy transferred (measured in joules).
"C" stands for electrical charge (measured in coulombs).

✦ Ohm's Law

$$V = I \cdot R$$

"V" stands for voltage (measured in volts).
"I" stands for current (measured in amps).
"R" stands for resistance (measured in ohms).

Activity Log

Activity	Date

What I did/saw/learned

Activity	Date

What I did/saw/learned

Activity	Date

What I did/saw/learned

Activity Log

Activity	Date
What I did/saw/learned	

Activity	Date
What I did/saw/learned	

Activity	Date
What I did/saw/learned	

Activity Log

Activity	Date
What I did/saw/learned	

Activity	Date
What I did/saw/learned	

Activity	Date
What I did/saw/learned	

Activity Log

Activity	Date
What I did/saw/learned	

Activity	Date
What I did/saw/learned	

Activity	Date
What I did/saw/learned	

Glossary

Physics Glossary

A

- **Absolute zero** – Theoretically, the lowest possible temperature or the point at which molecular motion virtually ceases to exist. (0 °K or -465.67 °F or – 273.15 °C)
- **Acceleration** – A change in an objects speed, direction, or both.
- **Acoustics** – The study of how sound travels in a given space.
- **Actuator** – A piece of equipment that makes a robot move.
- **Air resistance** – The force that air exerts on an object as it falls.
- **Amplitude** – The size of a vibration or the height of a wave.
- **Anode** – A positively charged diode.
- **Antinoise** – Produced when two sound waves overlap and cancel each other out.
- **Arch** – A curved weight-bearing structure that is in the shape of an upside-down U.
- **Automaton** – A machine that can move by itself.

B

- **Balance** – A state of equilibrium when the forces acting on an object cancel each other out; also known as a zero resultant force.
- **Beam** – A horizontal, weight-bearing, rigid structure that carries a load.
- **Bridge** – A man-made structure that has been built to span rivers, canyons, roads, railways, and more.

C

- **Capacitor** – An electrical component that stores electrical charge and releases the charge when it is needed.
- **Cathode** – A negatively charged diode.
- **Concave** – A lens that curves inward.
- **Conduction** – The movement of electricity or heat through a substance.
- **Conductor** – A material through which electrical charge can easily flow.
- **Controller** – A computer or switch that can react to the information gathered by sensors.
- **Convection** – The movement within a fluid that is caused by the tendency of the hotter material to rise and the cooler material to sink, which results from the transfer of heat.
- **Converging lens** – A lens that causes parallel light rays passing through it to come together.

- **Convex** – A lens that curves outward.

D

- **Diffraction** – The spreading out of light waves when they pass through a narrow slit.
- **Diode** – An electrical component that allows electrical current to flow in and out in a single direction.
- **Diverging lens** – A lens that causes parallel light rays passing through it to spread out.

E

- **Echolocation** – The locating of objects through the use of reflected sound.
- **Effector** – A device that lets a robot affect things in the world around it, such as grippers, tools, and laser beams.
- **Electrical charge** – A charge produced by an excess or shortage of electrons.
- **Electrolyte** – A substance that can conduct electrical current.
- **Electromagnet** – A magnet that can be switched off and on with electric current.
- **Electromagnetic force** – The force produced when an electrical current flows through a wire; it forms a magnetic field.
- **Electromagnetic spectrum** – The full range of electromagnetic radiation, which includes radio waves, infrared rays, visible light, Ultraviolet light, X-rays, and Gamma rays.
- **Electromagnetic waves** – These waves carry energy from one place to another and are produced when an electric charge vibrates or accelerates.
- **Energy** – The ability to do work.
- **Energy conversion** – The process of changing one form of energy into another.
- **Engineer** – A person who uses science and math to design, build, and maintain engines, machines, bridges, tunnels, and other public works structures.
- **Entropy** – The degree of disorder in a given system.

F

- **Fission** – The process by which a heavy unstable nucleus is split into two lighter nuclei, which causes the release of several neutrons and large amounts of energy.
- **Fluid** – A substance that assumes the shape of the container it is in, such as a liquid or gas.
- **Force** – A push or pull that acts on an object.
- **Force field** – The area in which a force can be felt.
- **Frequency** – The number of waves that pass a given point in a second.

- **Friction** – A force that opposes the motion of objects that touch as they move past each other.
- **Fulcrum** – The point on which a lever rests or is supported and on which it pivots.
- **Fusion** – The collision and combination of two lighter nuclei to form a heavier, more stable nucleus, releasing a large amount of energy.

G

- **Gravity** – The force that acts between two masses; it is an attractive force.

H

- **Heat** – A form of energy that flows from a place of high temperature to a place of lower temperature.

I

- **Induction** – The transfer of electrical charge without contact between the materials.
- **Inertia** – The tendency of an object to resist a change in its motion.
- **Input force** – The force you put into a machine.
- **Insulator** – A material through which electrical charge cannot easily flow.
- **Interference** – The disturbance of a signal when two or more waves meet.
- **Internal combustion engine** – An engine that generates power for movement through the burning of a fuel, such as coal, gas, or oil, and air.

J

K

- **Kinetic energy** – The energy of an object in motion; it depends upon the object's mass and speed.

L

- **Lens** – A piece of transparent material with a curved edge that causes light to bend in a particular way.
- **Load** – An applied force or weight.

M

- **Magnetic field** – The area around a magnet in which the magnetic force can be felt.
- **Magnetic pole** – One of the two ends of a magnet where the force of attraction or repulsion is the strongest.
- **Mass** – The amount of matter in an object.
- **Mechanical wave** – A wave that travels through a medium, such as air, water, or solids.
- **Momentum** – The tendency of an object to keep moving until a force stops it.

N

- **Newton** – The measurement of force; 1 Newton (N) is the force is takes to move a one kilogram object at 1 meter per second squared ($1 \text{ N} = 1 \text{ kg} \cdot 1 \text{ m/s}^2$).
- **Non-renewable energy resources** – Energy sources that exist in limited quantities, such as oil and coal.
- **Nuclear reactor** – The part of a nuclear power plant where the nuclear fission reaction occurs.

O

- **Output force** – The force that a machine exerts on an object.

P

- **Parallel circuit** – An electrical circuit in which current can pass thought more than one path.
- **Photons** – Packets of electromagnetic energy.
- **Photoresistor** – A light sensor that is able to change the resistance in an electrical current depending upon the amount of light it receives.
- **Potential energy** – The energy that an object has stored; it depends upon the object's weight and height.
- **Pressure** – The amount of force pushing on a giving area.

Q

R

- **Radiation** – 1. The heat energy emitted by a solid object.

    2. A stream of particles from a source of radiation.
- **Radioactive decay** – The process by which a nucleus ejects particles by radiation until

stability is reached.

- **Radioisotope** – An unstable nucleus that has a different number of neutrons than a stable nucleus.
- **Reflection** – The bouncing back of light from a surface.
- **Refraction** – The change in direction of a light beam as it passes from one medium to another of different density.
- **Renewable energy resources** – Energy sources that can be replaced in a relatively short period of time, such as wind and solar.
- **Resistance** – The ability of a material to resist the flow of electrical current.
- **Resonate frequency** – The frequency at which an object naturally begins to vibrate.
- **Robot** – A machine that is able to sense, think, and act on its own.

S

- **Semiconductor** – A material that can act as a conductor or insulator based on its temperature.
- **Sensor** – A device in robotics that takes in information from the outside world.
- **Series circuit** – An electrical circuit in which current passes through components that are one after another.
- **Solenoid** – A coil of wire that behaves like a magnet when electric current passes through it.
- **Sound** – A mechanical wave, or vibration, that travels through a medium, such as air, and can be heard when it reaches a person's or animal's ear.
- **Speed** – The ratio of the distance an object moves to the amount of time the object moves.
- **Steam engine** – An engine that generates power for movement through the use of steam, which rapidly expands and condenses.

T

- **Temperature** – A measure of how much heat energy is present in a substance.
- **Terminal velocity** – The point at which the force acting on an object of air resistance is equal to the force of gravity acting on the object.
- **Truss** – A framework of beams of bars that can support structures, like bridges.
- **Tunnel** – A passageway that goes under or through a natural or man-made obstacle, such as rivers, mountains, building, roads, and more.
- **Turbine** – A machine with shafts and blades that are turned by the force of wind or steam; this movement generates energy that can be turned into electricity.

U

V

- **Velocity** – The speed of an object in a particular direction.
- **Vibration** – A quick back and forth movement, for example when sound waves causes a nearby glass of water to vibrate.

W

- **Wavelength** – The distance between the crest of one wave to the crest of another.
- **Weight** – The force with which an object's mass is pulled toward the center of the Earth.
- **Work** – The transfer of energy that occurs when a force moves or changes an object.

X

Y

Z